METHODS IN YEAST GENETICS AND GENOMICS

A Cold Spring Harbor Laboratory Course Manual

2015 Edition

RELATED TITLES FROM COLD SPRING HARBOR LABORATORY PRESS

Landmark Papers in Yeast Biology

The Early Days of Yeast Genetics

Yeast Intermediary Metabolism

Molecular Cloning: A Laboratory Manual, 4th ed

METHODS IN YEAST GENETICS AND GENOMICS

A Cold Spring Harbor Laboratory Course Manual

2015 Edition

Maitreya J. Dunham
University of Washington

Marc R. Gartenberg
Robert Wood Johnson Medical School
Rutgers, The State University of New Jersey

Grant W. Brown
University of Toronto

CSH PRESS COLD SPRING HARBOR LABORATORY PRESS
Cold Spring Harbor, New York • www.cshlpress.org

METHODS IN YEAST GENETICS AND GENOMICS
A Cold Spring Harbor Laboratory Course Manual, 2015 Edition

Publisher and Acquisition Editor	John Inglis
Director of Editorial Services	Jan Argentine
Project Manager	Inez Sialiano
Permissions Coordinator	Carol Brown
Director of Publication Services	Linda Sussman
Assistant Production Editor	Maria Ebbets
Production Manager	Denise Weiss
Cover Designer	Mike Albano

Front cover artwork: Genetic analysis of yeast. (*Top*) Ten yeast tetrads dissected on a plate. Each column of four colonies contains the products of a single meiosis. (*Bottom*) 1536 colonies from a synthetic genetic array analysis. Each colony is a haploid double-mutant segregant from a cross between a query mutant and strains from the nonessential gene deletion collection. Photo credits: (*Top*) Tina Sing, Department of Biochemistry, University of Toronto. (*Bottom*) Friederike Ewald and Kyle Mohler, Yeast Genetics and Genomics course 2014.

Library of Congress Cataloging-in-Publication Data

Dunham, Maitreya J.
    [Methods in yeast genetics]
    Methods in yeast genetics and genomics : a Cold Spring Harbor Laboratory course manual. – 2015 edition / Maitreya J. Dunham, University of Washington, Marc R. Gartenberg, Robert Wood Johnson Medical School, Rutgers, State University of New Jersey, Grant W. Brown, University of Toronto.
    pages cm
    Rev. ed. of: Methods in yeast genetics : a Cold Spring Harbor Laboratory course manual / David C. Amberg, Daniel Burke, Jeffrey Strathern (Cold Spring Harbor, N.Y. : Cold Spring Harbor Laboratory Press, 2005).
    Includes bibliographical references and index.
    ISBN 978-1-62182-134-2 (printed hard cover)
    1. Yeast fungi–Genetics–Laboratory manuals. I. Gartenberg, Marc R. II. Brown, Grant W. III. Title.
QK617.5.K35 2015
579.5'62–dc23

2015017092

For a complete catalog of all Cold Spring Harbor Laboratory Press publications, visit our website at www.cshlpress.org.

Contents

Preface   vii

Acknowledgments   viii

Yeast Strains   ix

Yeast Plasmids   xi

Yeast Nomenclature   xii

Plates and Media   xiv

A Few Words on Sterile Technique   xv

Yeast Resources   xvi

EXPERIMENTS

   I.   Transformation of Plasmids and Integrating DNAs   1

  II.   Looking at Yeast Cells   15

  III.   Manipulating Mating-Type and Epigenetic Transcriptional Silencing   29

  IV.   Mating, Meiosis, and Tetrad Dissection   43

  V.   Isolation and Characterization of Auxotrophic and Temperature-Sensitive Mutants   57

  VI.   Working with Essential Genes   69

  VII.   Synthetic Lethal Mutants and Random Sporulation   79

  VIII.   Measuring Mutation Rates and Studying Human Genetic Variation in Yeast   89

  IX.   Mutation Detection Using Comparative Genomic Hybridization   97

  X.   Mutation Detection Using Whole-Genome Sequencing and Linkage   105

  XI.   Barcode Sequencing and Comparative Functional Genomics   113

TECHNIQUES AND PROTOCOLS

  1.   Genomic Modifications with PCR Products   123

  2.   High-Efficiency Yeast Transformation   133

  3.   Using the C6 Cytometer   137

4. Modified Hoffman–Winston Genomic DNA Preparation   143

5. Indirect Immunofluorescence Microscopy   145

6. Yeast Vital Stains   151

7. Actin Staining in Fixed Yeast Cells   155

8. Preparation of Slides with Agarose Pads for Imaging of Live Yet Immobile Yeast   157

9. Sporulation and Tetrad Dissection   159

10. Making a Tetrad Dissection Needle   163

11. EMS Mutagenesis   165

12. Counting Yeast Cells with a Hemocytometer   167

13. Flow Cytometry of Yeast DNA Content Using SYBR Green   171

14. Storage and Handling of the Systematic Deletion Collection   175

15. Scoring SGA Screens with SGATools   179

16. Measuring DNA Concentration with the Qubit Fluorometer   189

17. Colony PCR   191

18. Training for the Plate Race   193

APPENDICES

Appendix A: Media   195

Rich Media, 195

Synthetic Media, 196

Sporulation Medium, 199

Indicator Media, 200

Drug Selection Media, 201

SGA Media, 207

Appendix B: Tetrad Dissection Sheets   211

Appendix C: 96-Well Plate Template   213

Appendix D: Templates for Making Streaks and Patches   215

Appendix E: General Safety and Hazardous Material Information   217

Index   227

Preface

The Yeast Genetics and Genomics course is a modern state-of-the-art laboratory course designed to teach the full repertoire of genetic approaches needed to dissect complex problems in the yeast *Saccharomyces cerevisiae*. The course is designed for researchers who wish to use budding yeast as a model to study multiple cellular processes. The curriculum for this course embodies three components: (1) a rigorous and comprehensive set of laboratory experiments; (2) a series of detailed theoretical lectures by the instructors describing current knowledge, concepts, techniques, and strategies used in yeast genetics and genomics research; and (3) a series of seminars by invited speakers describing their research, with an emphasis on unique and imaginative approaches at the forefront of the yeast genetics and genomics field.

The 11 experiments included here are designed to showcase a foundation of methods needed in any modern-day yeast laboratory. Combinations of classical and modern genetic approaches are emphasized, including the isolation and characterization of mutants, two-hybrid analysis, tetrad analysis, complementation, and recombination. Molecular genetic techniques, such as yeast transformation, mating-type switching, gene replacement by polymerase chain reaction (PCR), construction and analysis of gene fusions, and generation of mutations, are also covered. Experiments take advantage of the yeast gene deletion collection to identify various kinds of genetic interactions including a genome-wide synthetic genetic array screen. Additional experiments introduce fundamental techniques in yeast genomics, including both performance and interpretation of multiplexed sequencing and comparative genome hybridization to DNA arrays. Comparative genomics using different yeast strains is introduced as a powerful approach to studying natural variation, evolution, and quantitative traits. Modern cytological approaches are also core, such as epitope tagging and imaging yeast cells using green fluorescent protein (GFP)-protein fusions and a variety of fluorescent indicators for various subcellular organelles and transcriptional readouts. Overall, the goal of the experimental section is to provide sufficient experience to allow investigators to use the techniques in any laboratory. Please note that some methods have been condensed due to the time limitations of the course.

Acknowledgments

This laboratory course manual incorporates significant portions of the manuals used in previous Cold Spring Harbor Laboratory (CSHL) Yeast Genetics Courses. Although the experiments have undergone considerable revision and new techniques have been added, the basic structure of this Course has not changed since its beginning 45 years ago. We are indebted to our predecessors, Fred Sherman, Gerry Fink, Jim Hicks, Cal McLaughlin, Brian Cox, Mark Rose, Fred Winston, Phil Hieter, Susan Michaelis, Aaron Mitchell, Alison Adams, Chris Kaiser, Dan Gottschling, Tim Stearns, Dean Dawson, Orna Cohen-Fix, David Amberg, Dan Burke, Jeff Strathern, Jeff Smith, and Sue Jaspersen for their teaching and for making this course an important part of the yeast community. We also thank Will Ryu, Alison Gammie, and Greg Lang for their contributions to Experiment VIII, Tina Sing for her contributions to Experiment VI, and Elena Kuzmin and Charlie Boone for their contributions to Techniques and Protocols 15. We also thank the members of our labs who have contributed to development of protocols throughout the book. Special thanks to the many teaching assistants and Cold Spring Harbor Laboratory staff who have been essential to the course's success each summer.

We gratefully acknowledge the sources of funding and equipment support for the course over the years, including the National Science Foundation, Howard Hughes Medical Institute, the National Institutes of Health, Singer Instruments, Zeiss, BD, and Cold Spring Harbor Laboratory.

Yeast Strains

Saccharomyces cerevisiae was the first eukaryotic genome to be sequenced and annotated. The laboratory strains that most of us use are inbred from *Saccharomyces* strains from around the world. There are different types of laboratory strains designed for distinct purposes: the yeast deletion collection and the GFP library were created in derivatives of S288c (BY4741, BY4742, and BY4743), meiosis experiments are typically done in the SK1 strain background, aging and cell cycle studies are performed in a variant of W303, flocculation and pseudohyphal growth analysis are conducted in Σ1278B, and D273-10B is the favorite of mitochondrial labs. The choice of strain is often based on the history of the lab and the type of question being asked. The default reference sequence shown in the *Saccharomyces cerevisiae* Genome Database (SGD; www.yeastgenome.org) is from S288c. The sequences of other commonly used yeast strains can also be found on SGD on a genome-wide or gene-by-gene basis.

It is important to note that polymorphisms within the genome give each strain background its distinctive properties. Much of the information about strain peculiarities has been collected by SGD (http://wiki.yeastgenome.org/index.php/Commonly_used_strains). Therefore, it is important to perform controls using isogenic strains (meaning strains that are derived from the same genetic background) and to be knowledgeable about the pedigree of the particular strains at hand. An electronic strain database is the most common method for recording the genotypes of strains used within a lab. This database can contain information not only regarding the strain genotype, but also notes about its construction, growth requirements, and any other important data. Because digital records can be searched, this type of an archive will allow members of the lab to access information about a particular strain years after the creator has left the lab.

A commonly asked question of new yeast researchers is "How do I pick a strain background?" Often, the choice in many laboratories will be historical. It is easiest and most convenient to use the strain background that the rest of the local community uses since many mutants and epitope tagged alleles are readily available. Convenience is the reason why many labs choose to work in the BY strain background, because knock-outs and green fluorescent protein (GFP) and tandem affinity purification (TAP)-tagged alleles for most genes can be purchased individually or as a collection from GE LifeSciences, EUROSCARF, transOMIC Technologies, ATCC, and other vendors. It is important to note that the BY strain background contains complete deletions of most of the auxotrophic markers. Although this facilitates certain types of manipulations since it virtually eliminates unwanted recombination events, it is impossible to target plasmids into these same loci because there is no sequence homology. Prototrophic versions of S288c, including complete nonessential gene deletion collections and a collection of the titratable-promoter essential alleles, have recently been published by Amy Caudy and Markus Ralser and are available from EUROSCARF.

We are fortunate that the yeast community generously shares published reagents. Therefore, it is likely that in a career as a yeast geneticist, a researcher will both receive strains from other labs as well as be asked to share materials with other labs. The simplest way to do this is to use sterile foil packages that contain squares or disks of filter paper onto which a suspension of yeast has been spotted. These packets can be sealed and taped to a piece of paper for delivery through the mail system. Upon receipt, sterile forceps are then used to unwrap each foil packet and the cells are spread onto suitable media by dragging the filter paper over the surface of the plate. Another favorite method is using slants, which are simply small test tubes containing agar onto which the strain is streaked. Sending and receiving larger collections of yeast require additional steps, which is discussed in Techniques and Protocols 14, Storing and Handling the Systematic Deletion Collection.

Yeast strains can be preserved for decades, essentially in their original form. Preparation and use of frozen stocks of yeast will help ensure that strains remain intact and do not contain unwanted mutations that naturally occur with repetitive passaging of cells. Frozen stocks are most typically kept in a final glycerol concentration of 15%–20%, but DMSO, sorbitol, and trehalose are also sometimes used. Another important consideration in preserving pure strain lineages is to be aware of what crosses have been done to generate a strain. If an outcross has been performed, for example, to move a locus of interest from one background into a new strain background, the segregants will all be genetically different and are no longer isogenic. Even with repeated backcrossing, some of the outcross strain genome will remain (especially, of course, the part linked to the locus that came from the outcross strain). Such strains would be "congenic" to indicate that most of their genome is from a pure strain background, but with ambiguities. The Hieter series of strains (YPH499, YPH500, and strains derived from them) is one commonly used background that is congenic to S288c. A nonreverting *ura3* allele from the FL100 background was crossed into S288c and then backcrossed to S288c >10 times. Even with this repeated backcrossing, swaths of the FL100 genome remain, and YPH499 and YPH500, the *MAT**a*** and *MAT*α reference strains of this background, have different genome content. Background considerations are especially important for whole-genome sequencing analysis.

Yeast Plasmids

Many types of plasmids can be used for genetic analysis and for engineering of yeast strains. Many of these reagents can be purchased for a nominal fee from sources such as EUROSCARF, Addgene, and the ATCC. The Yeast Resource Center (http://depts.washington.edu/yeastrc/) also distributes plasmids that are useful for microscopy and for two-hybrid studies.

Often, the best (and cheapest) source of many reagents is the yeast lab down the hallway.

Yeast Nomenclature

As yeast geneticists, it is important that we share a common nomenclature to clearly communicate our findings to other scientists. The first type of nomenclature we need to master is classic nomenclature used by many geneticists. *Mutations* are perturbations in the DNA sequence that may or may not result in a *mutant* phenotype, that is to say an observable change when compared to the parent. Obviously, some mutations in open reading frames (ORFs) have no effect on an organism because they do not result in a change in protein sequence due to degeneracy in the DNA code. These are *silent mutations*. Other mutations that result in changes in a single amino acid are known as *missense mutations*. Mutations that result in truncations are called *nonsense mutations*. Still other mutations fall in noncoding DNA. The phenotype caused by mutations depends on multiple factors, and the classification of mutants is generally compared to the phenotype elicited by deletion of the gene, which can be done easily in yeast using molecular genetic methods to create a *null* allele. Partial loss-of-function alleles are common in yeast, particularly when studying essential genes. These are typically *hypomorphic* alleles. Gain-of-function mutants, or *hypermorphs*, can sometimes be caused by specific point mutations, or can be created by overproduction of a gene using a series of regulatable promoters or by placing the genes on high-copy-number plasmids.

The second type of important nomenclature is the nomenclature of *Saccharomyces cerevisiae*. The conventions of budding yeast nomenclature are different from that of fission yeast and other organisms. The nomenclature conventions are summarized on SGD (http://www.yeastgenome.org/help/community/nomenclature-conventions) and described briefly below.

- Yeast genes names are written with three letters and up to three numbers that are relevant to gene function, localization, or phenotype: *MPS1* (for monopolar spindle 1), *HSP12* (for heat shock protein 12).
- Wild-type genes are written with capital letters in italics: *ADE2, CDC15, MPS1*.
- Recessive mutant genes are written with small letters in italics: *ade2, cdc15, mps1*.
- Mutant alleles are designated with a dash and a number: *ade2-1, cdc15-2*.
- Where alleles were engineered at specific sites, an informative name about the nature of the mutation is written: *mps1-as1, mps1ΔA, mps3-A540D*.
- Gene deletions and the marker used for deletion are indicated: *mps1Δ::kanMX* (if there is no marker listed, it was most likely looped out).
- Dominant mutants are written in all capital letters and in italics with the allele number or mutation listed after: *CDC31-16, MPS3-G186K*.

- A gene product, a protein, is written with a capital letter at the beginning and is not in italics; often a "p" is added at the end, although this convention is falling out of favor in the literature: Ade2 or Ade2p, Cdc15 or Cdc15p, Mps1 and Mps1p.

- A mutant gene product is written in lowercase and not in italics: ade2-1 or ade2-1p, cdc15-2 or cdc15-2p, mps1-as1 or mps1-as1p.

- The exception are dominant alleles that are written with a capital letter at the beginning: Cdc31-16 or Cdc31-16p, Mps3-G186K or Mps3-G186Kp

- Systematic names for nuclear ORFs are given a landmark name: *YDR518C, YML016W...,* where
  Y stands for "yeast,"
  the second letter represents the chromosome (D = IV, M = XIII....),
  L or R indicates the left or right chromosome arm,
  the three-digit number stands for the ORF counted from the centromere on that chromosome arm (for genes discovered after the initial systematic names were assigned, a letter is appended: *YER078W-A),* and
  C or W stand for "Crick" or "Watson," i.e., indicate the strand or direction of the ORF.

- Some genes do not follow this nomenclature because they were named before the nomenclature conventions: *HO, MATa, MATα, RPL1A & RPL1B, OM45.*

- For information on the nomenclature of RNAs, mitochondrial genes, transposable elements, etc., see the nomenclature conventions on SGD.

It is also common to introduce DNA into yeast from non-yeast sources, in plasmids, and as epitope and fluorescent protein fusions using PCR-based methods. The nomenclature conventions for these types of DNA elements are not as standardized as the yeast genome naming. Three types of plasmids are used in yeast: integrating plasmids, plasmids that contain a centromere and an autonomously replicating sequence, and plasmids derived from the 2μ element. These can be abbreviated YIp, YCp and YEp for yeast integrating, centromeric, and episomal plasmid, respectively. More commonly, the type of plasmid, the marker used for plasmid selection, and information about the insert are listed. For example, p*CEN/ARS-URA3-MPS1* is a plasmid with a centromere, an autonomously replicating sequence, the *URA3* gene for plasmid selection, and the *MPS1* gene. When a gene is fused to an epitope or a fluorescent protein, the modified gene is listed as *SPC42-GFP* or *GFP-SPC42*, depending on the tag location relative to the gene. The resulting proteins are written using the same nomenclature listed for the genes: Spc42-GFP or GFP-Spc42. If a marker gene accompanies the addition of an epitope to an endogenous gene, the construction is specified: *SPC42-GFP::kanMX* or *kanMX::GFP-SPC42*. If integrated elsewhere in the genome, the site of integration is denoted: *trp1::SPC42-GFP::kanMX*.

Plates and Media

In general, yeast require a source of carbon, nitrogen, and other essential nutrients for growth in the lab. The specific requirements for growth will depend on the strain background. Most lab strains have been engineered to contain mutations in different nutritional markers to allow for molecular genetic manipulation. For strains to grow, these auxotrophies need to be complemented either by a wild-type version of the gene or by supplementation of the growth media. Auxotrophies are generally regarded as harmless, but be aware that some experiments may be affected by markers. Metabolism, gene expression, and other systems are perturbed by auxotrophies, even if the growth medium is supplemented.

In most labs, two types of media are commonly used: rich media and minimal media. YPD (or YEPD) is a rich media composed of yeast extract, peptone, and dextrose (glucose). In this media, a wild-type yeast strain will exhibit optimal growth and double approximately every 60–90 min at 30°C. Minimal media is composed of yeast nitrogen base (a defined mixture that contains sources of nitrogen, phosphate, sulfate, vitamins, trace elements, and salts) and dextrose, plus a defined set of amino acids and nucleotides as desired. Labs use different mixtures to make minimal media; however, the most common types are SC (Synthetic Complete) and CSM (Complete Supplement Mixture). If one nutrient is omitted from this powder, cells in which the auxotrophy is complemented can be selected. If leucine is omitted, the media would be called SC-Leu (synthetic complete minus leucine). Often, minimal media is referred to a drop-out media because of the nutrients that have been "dropped-out." Another common formulation is "add-back" medium, which starts with a synthetic defined base containing no amino acids or nucleotides, to which individual supplements are added back as required. Dominant drug-resistant genes from bacteria have been developed as markers for yeast and are now in common use. The drugs can be added to YPD or to minimal media that has been modified slightly. Sporulation also requires specialty media that is low in nitrogen and carbon. Recipes for these types of media and others are detailed in Appendix A.

A Few Words on Sterile Technique

Any lab space may harbor a diverse and potentially invasive microbiome. Fungal contamination in plates is particularly pernicious. Good sterile technique will help, although yeast geneticists still find themselves performing surgery on precious plates to cut out unwelcome fungal growths. If one of your plates is contaminated, please make sure to wrap it with Parafilm before disposing it in the biohazard trash. Spores spread easily. Keep your bench and other equipment clean by wiping them down with ethanol or diluted bleach, especially after a spill or after handling a contaminated sample. Wash your hands and/or change gloves often to avoid cross-contamination.

Work near an open flame, which creates updrafts that discourage dust from falling into your cultures. Before using a bottle of medium, gently swirl it and look for kicked-up cells that could indicate contamination. Particles in your media are not always contamination; sometimes they are just chemicals that did not dissolve. Knowing the difference comes with experience. Looking at a sample under the microscope can remove all doubt. Flame the lips of bottles before pipetting from them. Use serological pipettes to pipette from bottles and tubes. Using a Pipetman brings a risk of touching the sterile inside surfaces of a container with the nonsterile portions of the pipette and contaminating both sample and pipette. If a Pipetman is preferred, decant some of the media into a sterile smaller bottle or Falcon tube so the source will not become contaminated. Always run a media-only control when growing cultures, and remember to check that it is still clear after incubation. Do not contaminate your pipette tip. One easy way to do this is to turn a pipette on its side while it has culture in the tip, causing the liquid to run back toward the pipette shaft. Try to work quickly and efficiently so that samples are not left uncapped for long.

Yeast Resources

The best resource for *Saccharomyces cerevisiae* online is the *Saccharomyces* Genome Database (SGD) (http://www.yeastgenome.org). It contains detailed information on gene sequences, gene functions, protein localization, and genetic and physical interactions, which have been manually curated from peer-reviewed literature. It provides multiple tools to search for information, including data on genes, sequences, yeast labs, and more. Links to outside data sets, such as transcription profiles and GFP localization data, can also be accessed through SGD. Other useful sites for yeast include:

The Yeast Resource Center: http://depts.washington.edu/yeastrc/
BioGRID: http://thebiogrid.org
The Rothstein Lab: http://www.rothsteinlab.com/tools/
EUROSCARF: http://web.uni-frankfurt.de/fb15/mikro/euroscarf/
Princeton Yeast Functional Genomics Database: http://yfgdb.princeton.edu
SPELL: http://imperio.princeton.edu:3000/yeast
DRYGIN: http://drygin.ccbr.utoronto.ca/index.html
Yeast Papers Twitter: http://twitter.com/yeast_papers

These and many other sites can be accessed in the LOCUS Summary page for each gene in SGD.

Many currently available books describe the techniques and protocols used in yeast. SGD curates a list of many methods under the community tab. The most commonly used "bible" of yeast is "Guide to Yeast Genetics and Molecular Biology," (*Methods in Enzymology*, volumes 194, 351, 470) edited by Christine Guthrie, Gerry Fink, and Jonathan Weissman. These volumes include great introductions to methods, from starting a yeast lab to cell biology to systems biology. Two introductory articles on basic yeast methods can also be downloaded from Fred Sherman's website (http://www.urmc.rochester.edu/labs/Sherman-Lab/publications/books.cfm?redir=dbb.urmc.rochester.edu).

Information on specific biological processes in yeast have been summarized by experts in the field in an ongoing series of review articles published in *Genetics*. Known as *The Yeast Book*, each review can be accessed through the journal's website (http://www.genetics.org/site/misc/yeastbook.xhtml).

For more on the history of yeast genetics, James Barnett and co-authors have published a series of articles in the journal *Yeast*. Also, Cold Spring Harbor

Laboratory Press published *The Early Days of Yeast Genetics* edited by Michael Hall and Patrick Linder. The history of the Yeast Course is covered in a chapter by Gerry Fink in this book, in addition to a 2001 article in *Genetics* by Peter Sherwood. For a fictionalized account of the Yeast Course, read *The Marriage Plot* by Jeffrey Eugenides.

Transformation of Plasmids and Integrating DNAs

A great utility of *Saccharomyces cerevisiae* as a model organism is the ease with which it is transformed by exogenous DNA. Commonly used laboratory strains bear numerous auxotrophies that are complemented by commercially available shuttle vectors. In this section, we learn how to transform yeast with extrachromosomal plasmids.

An equally important utility of *S. cerevisiae* is the high frequency with which transforming DNAs undergo homologous recombination with the genome and with one another. It is reasonably simple to delete, modify, and replace the genes of yeast. Traditionally, these operations were performed with DNAs assembled by cloning in bacteria. Now, many genomic modifications are performed with strategically designed PCR (polymerase chain reaction) products. In this section, we also learn how to add, delete, and modify genes by homologous recombination with transforming DNAs.

TRANSFORMATION METHODS

The original method for yeast transformation involved incubating spheroplasted cells with DNA, polyethylene glycol (PEG), and $CaCl_2$ (Hinnen et al. 1978). A more convenient and much more widely used method involves the treatment of cells with the alkali salt lithium acetate (LiOAc), followed by incubation with DNA and PEG (Ito et al. 1983; Soni et al. 1993). DNA can also be introduced by electroporation, whereby a brief electrical pulse permeabilizes the cells to DNA, by agitation of cells with glass beads, by bombardment of cells with DNA-coated particles (currently the only way to transform mitochondria), and by direct conjugation between bacterial and yeast cells. The method of choice depends on the purpose of the experiment, the number of strains to be transformed, and the desired number of transformants. The efficiency of transformation is often the most important parameter. If the goal is simply to put a plasmid into a given strain (only a few colonies needed), then any of the methods will work; if the goal is to get 10^6

transformants for screening a library, then spheroplasting, LiOAc, and electroporation methods are preferable because each yields a high transformation frequency under optimal conditions.

SELECTABLE MARKERS

Although the transformation frequency of yeast can be quite high, only a small fraction of cells in an experiment can be transformed. Therefore, it is essential that transforming DNA be linked to a marker gene to select for transformed cells. Traditionally, selectable markers were chosen from biosynthetic pathways. Common markers include *LEU2*, *HIS3*, *ADE2*, and *TRP1*, which are required for the production of leucine, histidine, adenine, and tryptophan, respectively. Of equal importance to the choice of a selectable marker is the corresponding chromosomal mutation that causes the auxotrophy. This mutation should be completely recessive and preferably nonreverting; for example, the *leu2-3,112* allele carries a pair of frameshift mutations that revert with a very low frequency ($<10^{-10}$) during normal growth. Nevertheless, the allele can occasionally revert to *LEU2* via homologous recombination with transforming DNA bearing the wild-type gene. Consequently, orthologous biosynthetic genes from distant yeast have been used as markers in *S. cerevisiae*. For example, the *Schizosaccharomyces pombe HIS5* gene complements the *S. cerevisiae his3-11,15* mutation. More recently, dominant drug resistance has been used as a selectable phenotype in yeast (Hadfield et al. 1990; Wach 1996; Goldstein and McCusker 1999). The bacterial *kan^r*, *hph^r*, and *nat^r* genes confer resistance to G418 (a kanamycin derivative), hygromycin, and nourseothricin (ClonNat). These genes conveniently possess no sequence homology with the yeast genome. For a number of yeast marker genes, counterselections exist. In these cases, growth in the presence of a small molecule causes toxicity or death when the marker gene is present. For example, the enzyme encoded by the *URA3* gene converts 5-fluoro-orotic acid (5-FOA) into a toxic nucleic acid precursor (Boeke et al. 1984). Similar counterselection strategies exist for the common marker genes *TRP1* and *LYS2* using 5-fluoranthranilic acid and α-aminoadipate, respectively (Zaret and Sherman 1985; Toyn et al. 2000).

REPLICATION ORIGINS

DNAs taken up by yeast during transformation are maintained heritably only if they replicate, either autonomously as extrachromosomal plasmids or passively as integrated genomic elements. Extrachromosomal plasmids require a plasmid-borne origin of replication, known historically as an *ARS* (autonomous replicating sequence) (Struhl et al. 1979). *ARS* elements frequently correspond to genomic sites that act as chromosomal origins of replication. Because the *ARS* elements are relatively small and somewhat flexible in sequence, cloned DNA sequences from other organisms are occasionally found to possess *ARS* activity. Integrating plasmids cannot replicate on their own because they lack *ARS* elements. Indeed, *ARS* elements were first cloned by virtue of their conferring high transformation efficiency to integrating plasmids.

VECTOR SYSTEMS

The first *ARS*-bearing plasmids (ARS plasmids) segregated poorly during mitosis, yielding some daughter cells with no plasmids and others with numerous copies (Murray and Szostak 1983). ARS plasmids gain stability with the addition of a centromere (*CEN*), which attaches the plasmid to the mitotic spindle, ensuring segregation to both mother and daughter cells. Because of the high fidelity of segregation, *CEN/ARS* plasmids (CEN plasmid, for short) are typically maintained at one or two copies per cell during mitotic growth. In meiosis, CEN plasmids typically show a 2:2 segregation pattern (if the diploid cell had one copy) or 4:0 segregation pattern (if the diploid cell had at least two copies). CEN plasmid segregation is nevertheless imperfect. Plasmid-free cells are generated at a small but measurable rate. Consequently, CEN plasmid copy number can increase severalfold above one or two under conditions of strong selection for plasmid-encoded genes.

An autonomously replicating plasmid that has the 2μ origin of replication and partitioning locus (2μ-based plasmid) segregates as well as a CEN plasmid in mitosis. However, 2μ plasmids are present at much higher copy number, typically 20–50 copies per cell. The high fidelity of 2μ plasmid segregation depends on the presence of the endogenous 2μ plasmid, which encodes partitioning machinery utilized by all 2μ-based plasmids. If a strain lacks the endogenous 2μ plasmid, then 2μ-based plasmids segregate as ARS plasmids (Murray and Szostak 1983). Many older yeast plasmids were named systematically. In this system, integrating plasmids were designated YIp; ARS plasmids, YRp; CEN plasmids, YCp; and 2μ plasmids, YEp (E for episome). YRp plasmids are rarely used because of their extreme instability.

In addition to the yeast-specific elements, all standard yeast vectors also have a bacterial origin of replication and a bacterial selectable marker, usually ampicillin resistance. Older vectors (YIp5, YEp24, YCp50) are usually based on a pBR322 backbone. This allows the plasmids to be shuttled back and forth between bacteria and yeast. More recent plasmids contain (1) plasmid backbones that replicate to higher copy number in bacteria, (2) multiple cloning sites (MCSs), and (3) single-stranded phage origins for the isolation of DNA for sequencing. The well-designed and popular pRS series of yeast vectors by Sikorski and Hieter (1989) contain a common backbone yet differ by virtue of the yeast marker gene (*URA3*, *HIS3*, *LEU2*, and *TRP1*) and whether they replicate in yeast at low copy or high copy or as an

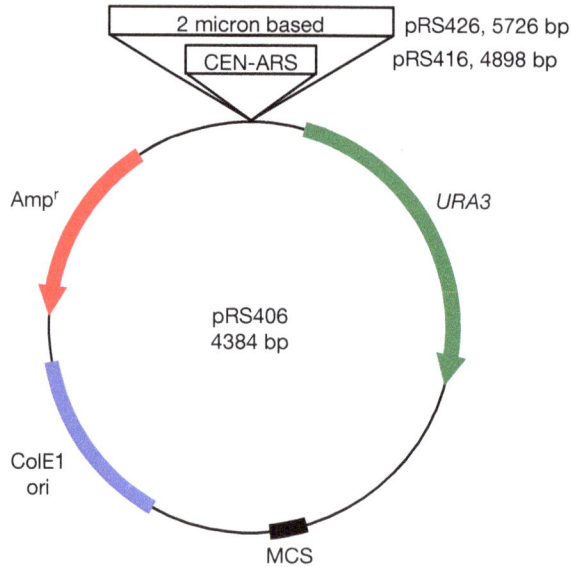

FIGURE 1. *URA3* members of the pRS series of plasmids.

integrated genomic element. Figure 1 illustrates the *URA3* plasmids of this series. pRS406 is an integrating vector, pRS416 is a low-copy CEN vector, and pRS426 is a 2μ-based vector. The shared structure of these plasmids facilitates convenient interchange of sequences between them by traditional cloning or by in vivo recombination methods in yeast.

INTEGRATION

Targeted integration of a plasmid into the yeast genome occurs by homologous recombination, which is stimulated roughly 100-fold by a double-strand break in the plasmid within the region of homology (Orr-Weaver et al. 1981). Integration by a single crossover yields a direct repeat of the homologous sequence, as shown in Figure 2A. Note that the entire plasmid is integrated, including the bacterial sequences and the yeast selectable marker. These now serve as a physical and phenotypic marker at the site of integration.

The integrating plasmid used in this experiment, pRS406-Nup49-GFP, contains *URA3* and a chromosomal fragment bearing the yeast *NUP49* gene. Thus, the plasmid can be directed to integrate at *ura3-1* in the strain used here or at the *NUP49* locus, depending on where a double-strand break is made by restriction enzyme digestion. Cutting the plasmid within *NUP49* results in nearly 100% integration at *NUP49* (Fig. 2B).

Integration at a specific locus can be verified both genetically and physically. The example in Figure 2A shows primers that can be used with PCR for

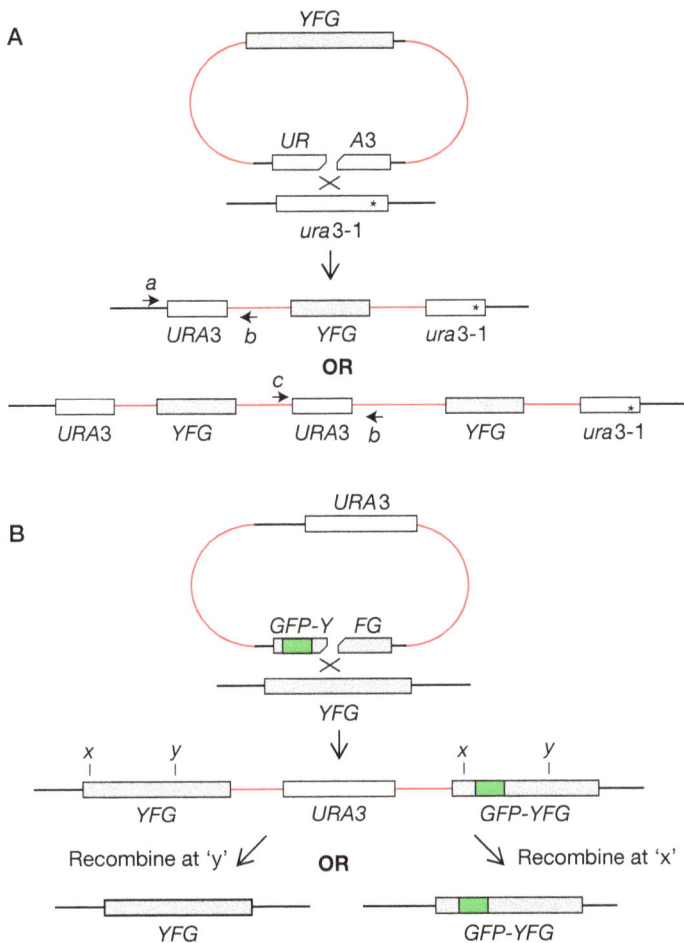

FIGURE 2. Genomic modification with integrating plasmids. (*A*) Single and multiple integration events are shown. Primer binding sites (*a, b,* and *c*) for analysis by PCR are listed. (*B*) Pop-in/ Pop-out strategy for chromosomal gene modification. Representative crossover sites for reversal of integration are shown (labeled with *x* and *y*).

verification—primer "*b*" binds to a unique sequence in the integrated plasmid, whereas primer "*a*" binds to a chromosomal site adjacent to the point of integration. Sometimes, sequential integration events occur within the same cell to yield multimers of the integrated plasmid. Additional PCRs can be used to identify the existence of multimeric inserts (Fig. 2A, primers *b* and *c*).

The recombination events that yield integration can also spontaneously reverse, resulting in eviction of the integrated DNA. These events are rare (one in 10^4 cells) and are usually not problematic for typical experiments. In certain circumstances, however, reversal of integration is desirable. In the Pop-in/Pop-out technique, a *URA3* plasmid is first used to modify a gene through the addition of sequences by homologous recombination. In the example in Figure 2B, a GFP (green fluorescent protein) module is added to a chromosomal gene. 5-FOA is then used to select for reversal of the integration, which eliminates the plasmid sequences and either leaves behind the GFP modification or restores the gene to its original state, depending on the site of crossover (between the sites labeled "*x*" or those labeled "*y*" in Fig. 2B).

PCR-MEDIATED GENOME MODIFICATIONS

Numerous modifications of the yeast genome are now performed with PCR products rather than engineered plasmids. This approach bypasses traditional cloning techniques that are both time consuming and limited by available restriction digestion sites. Using appropriately designed PCR products, it is possible to delete genes, swap alleles, add epitope tags, and integrate foreign DNA. As with the integration of plasmids, integration of PCR products requires sequence homology at the ends of the integrating DNA. Unlike plasmids, the flanking regions of PCR homology are typically orientated in the "omega" or "ends-out" configuration that requires two separate crossover events (Fig. 3). Homologous regions are appended onto PCR products using primers that contain 40–60 bp of homologous DNA (see Techniques and Protocols 1). Integration is enhanced with additional DNA homology, which can be obtained with a second round of PCR and new primers or by stitching together PCR products (Reid et al. 2002).

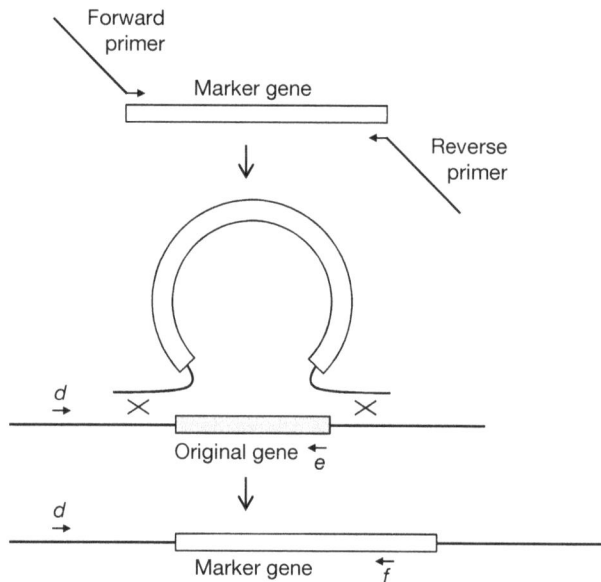

FIGURE 3. PCR-mediated gene replacement. A marker gene is amplified with primers that add 40–60 bp of homology for recombination via the ends-out orientation. Binding sites for primers to confirm gene replacement (*d*, *e*, and *f*) are shown.

Numerous collections of yeast have been assembled that contain systematic modifications of nearly every yeast open reading frame (ORF), each with a tightly linked marker gene. Notable examples include a complete gene knockout collection (available as heterozygous diploids), a collection of ORFs tagged with GFP, and a collection of ORFs tagged with the TAP tandem-affinity reagent. The collections are particularly useful in PCR-mediated gene modification approaches because they provide ready-made templates for PCR. In a typical application, a genomic region is amplified using short primers that capture the feature of interest, as well as the marker gene and 200 bp of flanking homologous sequence. The increased size of the homologous domains increases the efficiency and specificity of genomic integrations in strains relevant to the experiment at hand.

STRAINS

1-1 HJ1 *MAT**a** ade2-1 can1-100 his3-11,15 leu2-3,112 trp1-1 ura3-1 Δsir2::natMX*

PLASMIDS

pRS416
pRS416-GPDp-GFP
pRS406-GPDp-GFP
pRS406-NUP49-GFP

PRIMERS

f150—5′-TTCGAGTACGGAGGGTATGGT-3′
r250—5′-CGGACATTTTCATTCAAGAGG-3′
d—5′-GCACCATTGTCAACGATATAAAC-3′
e—5′-GGTCAACTCGTCCTCATCTTCA-3′
f—5′-ACTCGCATCAACCAAACC-3′

EXPERIMENT OVERVIEW

Experiment IA

We will transform yeast using plasmids that integrate into the genome or replicate autonomously. The plasmids carry a *URA*3 selectable marker, as well as a GFP expression cassette that will be used to characterize the transformants. Our examination highlights the differences in copy number and stability provided by integrating versus extrachromosomal CEN/ARS vectors.

Experiment IB

We will transform a linearized plasmid that integrates at the *NUP*49 gene and tags the encoded nucleoporin with GFP. A Pop-in/Pop-out procedure will be used to eliminate plasmid sequences while leaving GFP in the chromosome. The experiment illustrates how alleles can be swapped and how crossover location is important when dealing with reversible modifications of the genome.

Experiment IC

We will knock out a gene by one-step gene replacement using a preexisting mutant as a template DNA for PCR. We will distinguish between gene replacement and off-target events by phenotypic analysis and PCR.

EXPERIMENTAL PROCEDURES

▶ Day 1

Experiments IA, IB, and IC

Restreak strain 1-1 on a YPD plate. Incubate at 30°C.

▶ Day 2

Experiment IC

You will be given a yeast genomic DNA extract that contains *hst1::kanMX*. Use PCR to amplify the *hst1::kanMX* domain with primers f150 and r250 according to Techniques and Protocols 1 (Genomic Modifications with PCR Products) (1905 bp). Perform this step in conjunction with PCR for Experiment II.

▶ Day 3

Experiment IA

In the morning, inoculate 5 mL of YPD with a single colony of strain 1-1. Incubate at 30°C overnight with agitation.

Experiment IC

Confirm PCR product by agarose gel electrophoresis. Perform this step in conjunction with gel for Experiment II.

▶ Day 4

Experiments IA, IB, and IC

In the morning, dilute overnight culture of strain 1-1 to 0.1 OD in 100 ml and grow to 0.3–0.5 OD. You will be given aliquots of the following plasmids for transformation:

1. pRS416

2. pRS416-GPDp-GFP

3. pRS406-GPDp-GFP cut with *Stu*I

4. pRS406-Nup49-GFP cut with *Xba*I

Harvest the cells by centrifugation at 2000 rpm in a clinical centrifuge and follow the protocol for LiOAc transformation (Techniques and Protocols 2, High-Efficiency Yeast Transformation). Transform strain 1-1 with the four DNA samples provided, as well as the *hst1::kanMX* PCR product from Day 2. Include a no-DNA control transformation. Plate the *URA3* plasmid transformants and half of the no-DNA control on SC-ura plates. Plate the *kanMX* transformant and other half of the no-DNA control on YPD plates. Autonomously replicating plasmids transform more efficiently than integrating vectors. Thus, spread only 20% of the reactions that used the pRS416 vectors. Place all plates in the 30°C incubator.

▶ Day 5

Experiment IC

Replica-plate your *hst1::kanMX* transformants from YPD to YPD + G418 plates. Incubate at 30°C.

▶ Day 6

Experiments IA and IB

Restreak several URA + transformants for single colonies on SC-ura plates. This step separates true transformants from neighboring untransformed cells.

▶ Day 7

Experiment IC

Replica-plate the *hst1::kanMX* transformants from the G418 plate to a nourseothricin (Nat) plate.

▶ *Day 8*

Experiment IA

Inoculate one transformant each of pRS416, pRS416-GPDp-GFP, and linearized pRS406-GPDp-GFP in 5 mL of SC-ura, +ade. Incubate in a shaker at 30°C overnight. These cultures will be used to evaluate the GFP levels. The extra adenine suppresses accumulation of red fluorescent pigment that interferes with cytometry.

Experiment IB

Confirm the Nup49-GFP transformants by fluorescence microscopy. Superior images are obtained with cells from mid-log cultures. For brevity, however, we will examine cells directly from the selection plate. Use a sterile pipette tip to dab a small amount of cells into 4 µL of H_2O on a glass microscope slide. Add a coverslip and seal with nail polish.

▶ *Day 9*

Experiment IA

Use the C6 cytometer to evaluate GFP levels in each of your overnight cultures (Techniques and Protocols 3, Using the C6 Cytometer). Note in your lab book the differences in fluorescence distributions. Why do the different plasmids yield different flow cytometry profiles?

If desired, these strains can also be examined by fluorescence microscopy. Concentrate cells and add 3–4 µL of cell suspension to a slide. Affix a coverslip with nail polish. How does the fluorescence vary from cell to cell for each sample?

▶ *Day 10*

Experiment IB

To select for Pop-outs of pRS406-Nup49-GFP, inoculate a transformant in 1 mL of YPDA (yeast peptone dextrose adenine) for overnight growth at 30°C.

Experiment IC

Inoculate a colony that is both G418^r and NAT^r in 5 mL of YPDA for overnight growth at 30°C. Inoculate strain 1-1 as a control.

▶ *Day 11*

Experiment IB

Spread 250 µL of the pRS406-NUP49-GFP overnight culture on a 5-FOA plate.

Experiment IC

From 1 mL of *hst1::kanMX* culture, isolate DNA according to Techniques and Protocols 4 (Modified Hoffman–Winston Genomic DNA Preparation) (stop at Step 11 and resuspend in 500 µL of H_2O). Use 1 µL of template to carry out PCR with primer combination d and f to test for *hst1::kanMX* knockout at proper location (1926 bp). In addition, perform PCR using primer combination d and e to test for loss of original *HST1* locus (740 bp). Check products on agarose gel.

▶ Day 14

Experiment IB

Identify colonies that have lost the *URA3* marker. Use a sterile pipette tip to dab small amounts of individual colonies into 4 µL of H_2O on glass microscope slides. Include the original URA+ transformant as a control. Do all of the isolates yield Nup49-GFP fluorescence? In the fluorescent cells, how does the intensity compare to the initial URA+ transformant?

MATERIALS

Day 1	1 YPD plate
	Strain 1-1
Day 2	DNA prep of strain JXY5 bearing Δ*hst1::kanMX4*
	PCR reagents (see Techniques and Protocols 1)
	Primers f150 and r250
	Q5 DNA polymerase
	5× PCR buffer
	10 mm dNTPs
Day 3	1 Culture tube with 5 mL of YPD
	Gel electrophoresis reagents (see Techniques and Protocols 1)
	Ultrapure agarose
	1× TBE
	1-kb ladder
	Loading dye
	SYBR Safe stain in trough
Day 4	500-mL Erlenmeyer flask containing 100 mL of YPD
	5 SC-ura plates
	2 YPD plates

DNAs for transformation:
1. pRS416
2. pRS416-GDPp-GFP
3. pRS406-GPDp-GFP cut with *Stu*I
4. pRS406-Nup49-GFP cut with *Xba*I

LiOAc transformation reagents (see Techniques and Protocols 2)
LiOAc in TE
Salmon sperm DNA
PEG solution
DMSO
TE
Sterile glass beads

Day 5   2 YPD + G418 plates
Velvets

Day 6   4 SC-ura plates

Day 7   1 Nat plate
Velvets

Day 8   3 Culture tubes with 5 mL of SC-ura, +ade
Glass slides, coverslips, and nail polish

Day 10   3 Culture tubes containing 5 mL of YPD

Day 11   SC + FOA plate
DNA isolation reagents (see Techniques and Protocols 4)
Glass beads
Lysis buffer
Phenol:chloroform
Ethanol
PCR reagents (see Techniques and Protocols 1)
Primers d, e, and f
Q5 DNA polymerase
5× PCR buffer
10 mM dNTPs

Gel electrophoresis reagents (see Techniques and Protocols 1)
    Ultrapure agarose
    1× TBE
    1-kb ladder
    Loading dye
SYBR Safe stain in trough

Day 14   Glass slides, coverslips, nail polish

REFERENCES

Boeke JD, LaCroute F, Fink GR. 1984. A positive selection for mutants lacking orotidine-5′-phosphate decarboxylase activity in yeast: 5-Fluoro-orotic acid resistance. *Mol Gen Genet* **197**: 345–346.

Goldstein AL, McCusker JH. 1999. Three dominant drug resistance cassettes for gene disruption in *Saccharomyces cerevisiae*. *Yeast* **15**: 1541–1553.

Hadfield C, Jordan BE, Mount RC, Pretorius GHJ, Burak E. 1990. G418-resistance as a dominant marker and reporter for gene expression in *Saccharomyces cerevisiae*. *Curr. Genet.* **18**: 303–314.

Hinnen A, Hicks JB, Fink GR. 1978. Transformation of yeast. *Proc Natl Acad Sci* **75**: 1929–1933.

Ito H, Fukuda Y, Murata K, Kimura A. 1983. Transformation of intact yeast cells treated with alkali cations. *J. Bacteriol* **153**: 163–168.

Murray AW, Szostak JW. 1983. Pedigree analysis of plasmid segregation in yeast. *Cell* **34**: 961–970.

Orr-Weaver T, Szostak J, Rothstein R. 1981. Yeast transformation: A model system for the study of recombination. *Proc Nat Acad Sci* **78**: 6354–6358.

Reid RJ, Lisby M, Rothstein R. 2002. Cloning-free genome alterations in *Saccharomyces cerevisiae* using adaptamer-mediated PCR. *Methods Enzymol* **350**: 258–277.

Sikorski RS, Hieter P. 1989. A system of shuttle vectors and yeast host strains designed for efficient manipulation of DNA in *Saccharomyces cerevisiae*. *Genetics* **122**: 19–27.

Soni R, Carmichael JP, Murray JA, 1993. Parameters affecting lithium acetate-mediated transformation of *Saccharomyces cerevisiae* and development of a rapid and simplified procedure. *Curr Genet* **24**: 455–459.

Struhl K, Stinchcomb DT, Scherer S, Davis RW. 1979. High-frequency transformation of yeast: Autonomous replication of hybrid DNA molecules. *Proc Natl Acad Sci* **76**: 1035–1039.

Toyn JH, Gunyuzlu PL, White WH, Thompson LA, Hollis GF. 2000. A counterselection for the tryptophan pathway in yeast: 5-Fluoroanthranilic acid resistance. *Yeast* **16**: 553–560.

Wach A. 1996. PCR-synthesis of marker cassettes with long flanking homology regions for gene disruptions in *S. cerevisiae*. *Yeast* **12**: 259–265.

Zaret KS, Sherman F. 1985. α-aminoadipate as a primary nitrogen source for *Saccharomyces cerevisiae* mutants. *J Bacteriol* **162**: 579–583.

Looking at Yeast Cells

At a diameter of about 5 µm, haploid yeast cells are large enough that cellular structures can be characterized by the light microscope. In this section, we explore the ways in which light microscopy is commonly used in yeast research. First, we become familiar with the shapes of yeast cells in various stages of development. Next, we use fluorescent dyes and drugs to illuminate various cell structures. Finally, we make fluorescent protein chimeras to visualize specific cellular structures. This technique is used to monitor aspects of mitotic chromosome segregation in living cells and in mutants that arrest during cell cycle progression.

EXAMINATION OF GROWTH PROPERTIES

Saccharomyces cerevisiae cells grow by budding. A cell that gives rise to a bud is called a mother cell, and the bud is referred to as the daughter cell. A new bud emerges from a mother cell close to the beginning of the cell cycle and continues to grow until it separates from the mother cell at the end of the cell cycle. Because all of the growth of a yeast cell is concentrated in the bud, and because this growth is essentially continuous, the size of the bud gives an approximate indication of the position of a given cell in the cell cycle. In an exponentially growing culture of yeast, about one-third of the cells are unbudded, one-third of the cells have a small bud, and one-third of the cells have a large bud (Figs. 1A and 1B). When cells in a culture consume all of the nutrients, they stop growing by arresting in the cell cycle without buds (Fig. 1C). Thus, a simple way of determining the growth state of a culture is to determine the fraction of budded cells with a microscope. Note that for some strains, the mother and daughter cells remain stuck together even after cytokinesis. Sonication is sometimes needed to separate these cells prior to microscopy. It is good laboratory practice to routinely examine cultures by phase-contrast microscopy to assess the physiological state of the cells and to detect contamination.

Many mutants arrest growth in ways that are diagnostic of the underlying molecular defect. For example, cells with improperly assembled mitotic spindles arrest with large buds, a point in the cycle that would normally correspond to

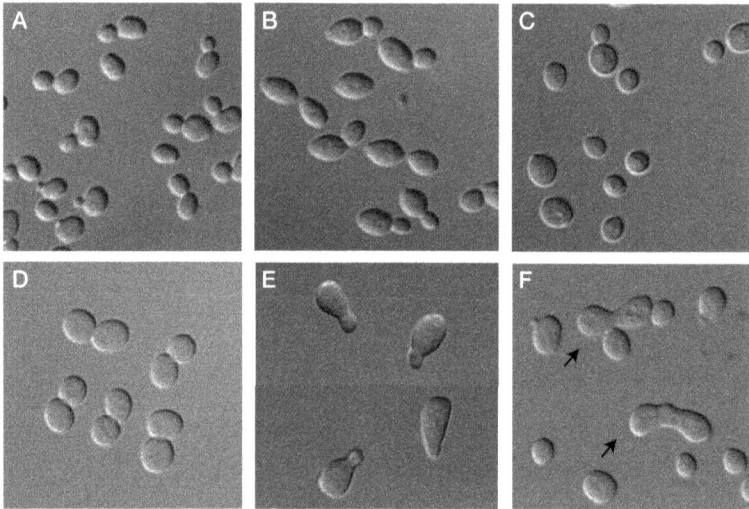

FIGURE 1. Cell morphologies of budding yeast at 100× magnification. (*A*) Asynchronous population of haploids. (*B*) Asynchronous population of diploids. (*C*) Starved haploids. (*D*) *MAT***a** haploids arrested in M phase with a nocodazole, a microtubule depolymerizing agent. (*E*) *MAT***a** haploids arrested in G_1 phase by mating pheromone. Note the distinctive "shmoo" shape. (*F*) Zygotes (marked with arrows) among haploids.

mitosis (Hartwell et al. 1973). It is important to note that the arrest point, or terminal phenotype, of mutant cells can be morphologically distinct from any cell type seen in a normal culture. For instance, when defects in the mitotic spindle arise mother and daughter cells continue to grow at the arrest point until both are much larger than normal yeast cells (Fig. 1D).

HAPLOIDS VS. DIPLOIDS

Although haploid and diploid yeast cells are morphologically similar, they differ in several important ways. Diploid cells are larger and more ovoid than haploid cells. The longest dimension of a diploid cell is about 1.3 times that of a haploid cell. This difference can be seen when haploids and diploids are compared side by side (compare Fig. 1,A and B). Because they are larger, diploid cells (or even tetraploids in some cases) are sometimes used in fluorescence microscopy where the larger size helps to resolve small cellular structures. Diploids can also be distinguished from haploids by their distinct pattern of budding. Yeast cells generally bud about 20 times before becoming senescent. Haploids bud axially with each bud emerging adjacent to the site of the previous bud (most W303-derived strains are exceptions). Diploids bud in a polar pattern with successive buds emerging from either end of the elongated mother cell.

MATING CELLS

Haploid yeast exist as either one of two mating types, *MAT*a and *MAT*α, that mate with each other to form *MAT*a/*MAT*α diploids. The mating process between two cells begins with an exchange of pheromones that induces the expression of proteins required for mating and causes the cells to arrest in G_1 phase without buds. Pheromones also cause the cells to form a projection from the cell surface for cell fusion (Fig. 1E). On solid media, this projection orients toward the mating partner. A cell with a mating projection is called a "shmoo." Shmooing cells join at the tips of their projections and their cytoplasms fuse, followed by fusion of their nuclei in a process termed karyogamy. The newly formed diploid is termed a zygote that is particularly easy to identify when the first bud emerges to create a trilobed shape (Fig. 1F). It is possible to isolate zygotes by micromanipulation (as in Experiment IV), allowing for the isolation of diploid cells, even in situations where no genetic selection for diploids exists.

STAINING CELLS WITH DYES AND DRUGS

Vital dyes are small fluorescent compounds that stain specific objects and organelles in live cells. They provide a convenient and rapid way to assess the location and behavior of organelle of interest, particularly when many different conditions and strains or mutants must be tested. When working with vital dyes, always use the lowest concentrations possible to avoid perturbations caused by the binding of dyes. Some dyes fluoresce only when bound to target. Other dyes fluoresce independent of context. In these cases, it is best to wash off unbound dye to increase signal over background. The following sections briefly describe the dyes and drugs we use in this course.

Calcofluor

Calcofluor, a brightener found in clothes detergents, stains the chitin ring that forms at sites of bud emergence. These rings persist at the septum between dividing cells and remain on the mother as "bud scars" after the daughter cell has departed (Fig. 2A). The number of bud scars on a mother cell is equal to the number of daughters she has produced, and thus bud scars serve as a metric of replicative age. By illuminating the distribution of bud scars on a cell's surface, the drug reveals whether an older cell has budded in an axial or polar fashion.

DAPI

DAPI (4′,6-diamino-2-phenylindole) fluoresces brightly when it intercalates between the bases of the DNA double helix but not when it binds to RNA. Binding to chromosomal DNA yields a large spot of nuclear fluorescence, whereas binding to

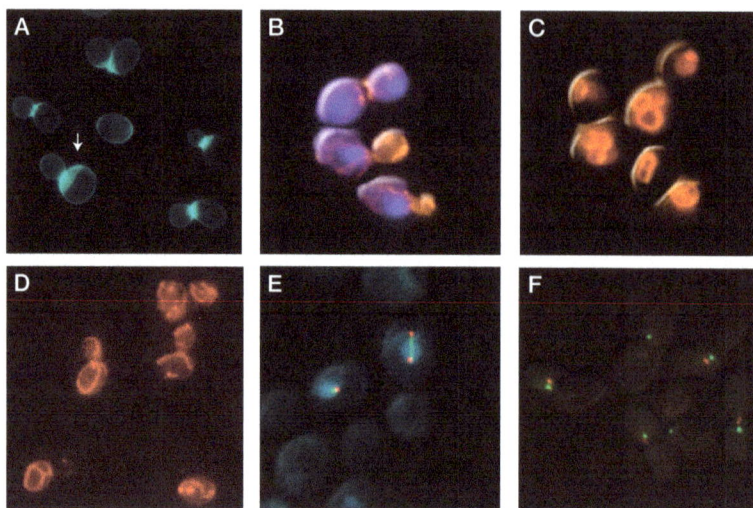

FIGURE 2. Imaging budding yeast with dyes, drugs and protein chimeras. (*A*) Calcofluor labeling of chitin rings. An arrow points to a cell with multiple axial bud scars. (*B*) DAPI labeling of DNA (blue) and rhodamine-phalloidin labeling of filamentous actin (orange). (*C*) FM 4-64 labeling of vacuoles. (*D*) Mitotracker Red labeling of mitochondria. (*E*) DNA stained with DAPI (blue) and mitotic spindle components visualized with Tub4-mCherry and GFP-Tub1 in fixed yeast. (*F*) Tub4-mCherry and Lac-GFP bound to a Lac operators on chromosome III in live yeast. Images were graciously provided by participants in the 2014 Yeast Genetics and Genomics Course: Valerie Thomas and Gustavo Silva (*B, C*); Nádia Maria Sampaio and Rhesa Ledbetter (*E*); Melinda Borrie (*A, D, E,* and *F*).

mitochrondrial DNA yields numerous smaller spots of cytoplasmic fluorescence (Fig. 2, B, D, and E). The compound can be used in live cells, but it stains more evenly when the cells have been fixed with ethanol. We will use DAPI in conjunction with rhodamine-phalloidin and with fluorescent protein chimeras.

FM 4-64

FM 4-64 is a lipophilic styryl dye that binds membranes and fluoresces within the hydrophobic membrane environment. The dye binds the plasma membrane first and then enters the cell by endocytosis. In pulse-chase experiments, it is possible to see the dye transfer from the plasma membrane to vesicles. Ultimately, FM 4-64 accumulates in vacuoles (Fig. 2C). The drug was particularly useful in screening for endocytic mutants (Vida and Emr 1995).

Mitochondrial Dyes

DiOC$_6$(3) (3,3'-dihexyloxacarbocyanine iodide), DiIC$_5$(3) (1,1'-dipentyl-3,3,3',3'-tetrame-thylindocarbocyanine iodide), and Mitotracker Red accumulate in all membranes but fluoresce most brightly in mitochondrial membranes due to the membrane potential (Fig. 2D). The dyes are particularly useful for examining mitochondrial morphology, which changes in response to environmental cues. At high levels, mitochondrial specificity of each of these dyes is lost, resulting in fluorescence of other cellular membranes, including the endoplasmic reticulum (ER). Of the three dyes, only Mitotracker Red remains in the mitochondria after fixation. Thus, Mitotracker Red is most appropriate if a secondary labeling procedure requires fixation.

Yeast cells with dysfunctional mitochondria cannot carry out oxidative phosphorylation and thus stain less brightly with these membrane-potential-dependent dyes. Such mutants grow more slowly than wild-type cells and are unable to grow on nonfermentable carbon sources such as lactate, glycerol, or ethanol. French scientists who first characterized mitochondrial mutants called this the "petite" phenotype. Petite strains form small milky-white colonies on fermentable carbon sources. This lack of pigmentation is most evident in an *ade2* background; formation of the red pigment that typifies *ade2* mutants requires oxidative phosphorylation. Diploid petite strains are also unable to sporulate, and it is wise to check a nonsporulating strain for the ability to grow on a nonfermentable carbon source before other potential causes of sporulation failure are examined. Petite mutants appear with high frequency in many common lab strains, in some cases, as many as 10% of the cells in culture. Although the petite phenotype can be due to mutations in either the mitochondrial or nuclear genomes, the great majority of petites are due to mitochondrial DNA mutations. The mitochondrial genome is given the designation "ρ." Wild-type strains are ρ^+; strains with deleted versions of the mitochondrial genome (the most common type of mutation) are ρ^-. Strains lacking the mitochondrial genome entirely are ρ^o. A common misconception is that ρ^o strains lack mitochondria altogether. Several essential reactions take place within the mitochondrial membrane, and even in ρ^o strains, a diminished mitochondrial structure can be seen in the electron microscope.

Rhodamine-Phalloidin

Phalloidin is a mushroom toxin that binds filamentous actin. When coupled to rhodamine, the drug offers a relatively simple way to assess actin structures within yeast. Rhodamine-phalloidin does not work with live yeast cells and is thus not a vital dye. After a simple fixation procedure, however, the reagent readily illuminates filamentous actin in patches, cables, and rings (Fig. 2B). These actin assemblies mediate cellular events such as budding, endocytosis, organelle trafficking, and cytokinesis. Microtubules, the other principal cytoskeletal component of budding yeast, mediate chromosome segregation and nuclear movement. Microtubules are best visualized with fluorescent protein chimeras and indirect immunofluorescence, as described in the following sections.

MICROSCOPY WITH FLUORESCENT PROTEIN CHIMERAS

Yeast have been used increasingly in cell biology, benefiting greatly from the development of robust methods to determine the intracellular location of gene products. Perhaps the most powerful and versatile method makes use of fluorescent proteins, such as green fluorescent protein (GFP), a naturally fluorescent protein from the jellyfish *Aequoria victoria*. GFP retains its fluorescence when expressed in bacterial, fungal, plant, and animal cells, making it an ideal fluorescent marker protein. Additional variants of GFP with distinct spectral properties, including yellow fluorescent protein (YFP) and cyan fluorescent protein (CFP), as well as a red fluorescent protein (RFP or variant mCherry) from the coral *Discosoma*, allow colocalization to be analyzed and are suited for advanced imaging techniques such as fluorescence resonance energy transfer (FRET). Note that the ethanol fixation step in many DAPI staining procedures eliminates fluorescence of the most common forms of GFP. In these situations, fix the cells first with formaldehyde to preserve the GFP signal and then treat with ethanol (see Techniques and Protocols 6, Yeast Vital Stains).

In yeast, a fluorescent protein is commonly appended to the ends of endogenous genes to visualize the resulting chimeric protein in living cells. In Figure 2E, the mitotic spindle is visualized in live cells with tubulin tagged with GFP (GFP-Tub1) and the spindle pole body tagged with mCherry (Tub4-mCherry). Each new chimeric protein must be tested for full functionality as the fluorescent tag can be disruptive, sometimes in ways that are difficult to detect. Jonathan Weissman and Erin O'Shea systematically fused GFP to 4159 yeast open reading frames (ORFs) and identified 22 specific subcellular structures (Huh et al. 2003). The association of uncharacterized gene products with these cellular structures led to important insights into the function of novel genes and a more thorough understanding of diverse cellular functions. Belmont and colleagues adapted fluorescent protein technology to visualize specific chromosome regions in yeast by integration of tandem repeats of the bacterial Lac repressor binding site into the genome and expression of the Lac repressor fused to GFP (Straight et al. 1996) (Fig. 2F). These innovations, when combined with the powerful genetic features of yeast, led to the discovery of genes involved in chromosome cohesion and condensation, spindle assembly, DNA-damage repair, telomere maintenance, and nuclear organization.

INDIRECT IMMUNOFLUORESCENCE

Immunofluorescence is a traditional method that uses antibodies to visualize locations of proteins and entire cellular structures. Typically, fixed and processed cells are incubated first with a primary antibody to the protein of interest and then with a fluorescently labeled secondary antibody directed against the constant region of the primary antibody. This two-layered indirect immunofluorescence approach confers a number of advantages: (1) multiple secondary antibodies can bind a single primary antibody, thus significantly increasing the intensity of the fluorescence signal; (2) the primary antibody need not be encumbered with fluorescent analogs, which might interfere with affinity or specificity; and (3) the experimenter can choose from a wide range of commercially available secondary antibodies that differ by the attached fluor and the animal species from which they came. Simultaneous labeling of multiple structures can be performed if different animals are used to generate the primary antibodies, and if these are matched with appropriately distinct secondary antibodies.

The requirement for a primary antibody can be both the central advantage and limitation of indirect immunofluorescence. In particular, an avid and selective antibody toward a protein of interest means that imaging can be done on panels of strains without having to make genetic modifications to accommodate imaging. In the absence of a good primary antibody, gene modification techniques can be used to add a short peptide epitope or an entire fluorescent protein. However, as stated above, protein functions can be altered by the modifications used to detect them. Another major disadvantage of immunofluorescence techniques is that cells must be fixed prior to observation. Thus, dynamic changes in structure or location cannot be assayed. In addition, the cell wall must be removed to allow access of the antibodies. Care must be taken to avoid artifacts due to cell preparation. Although not included in this year's set of experiments, indirect immunofluorescence remains a fundamental technique in the toolbox of yeast cell biologists. For a detailed procedure, see Techniques and Protocols 5 (Indirect Immunofluorescence Microscopy).

STRAINS

2-1	W303-1A	*MATa ade2-1 can1-100 his3-11,15 leu2-3,112 trp1-1 ura3-1*
2-2	W303-1B	*MATα ade2-1 can1-100 his3-11,15 leu2-3,112 trp1-1 ura3-1*
2-3	W303	*MATa/MATα ade2-1/ade2-1 can1-100/can1-100 his3-11, 15/his3-11,15 leu2-3,112/leu2-3,112 trp1-1/trp1-1 ura3-1/ura3-1*
2-4	BY4741	*MATa his3Δ1 leu2Δ0 met15Δ0 ura3Δ0*
2-5	BY4743	*MATa/MATα his3Δ1/his3Δ1 leu2Δ0/leu2Δ0 met15Δ0/MET15 LYS2/lys2Δ0 ura3Δ0/ura3Δ0*
2-6	MRG5379	*MATa bar1 ade2 ura3::GFP-TUB1-URA3 TUB4-mCherry:: kanMX TRP1*
2-7	MRG5652	*MATa cdc15-2 bar1 ade2 ura3::GFP-TUB1-URA3 TUB4-mCherry::kanMX TRP1*
2-8	MSB41	*MATa cdc31-2 bar1 ade2 TUB4-mCherry::kanMX TRP1*

2-9   MRG5651   *MAT**a** cdc16-1 bar1 ade2 ura3::GFP-TUB1-URA3*
                   TUB4-mCherry::kanMX TRP1

2-10   CSW91   *MAT**a** lys2::256xlac^{ops}-TRP1 ade2-1::HIS3p-GFP-lacI::HIS3*

PLASMIDS

pFA6a-link-mCherry

PRIMERS

Tub4-F5: 5′-TTGGAAGAGGACCTGGATGCCGACGGTGATCATAAATTAGTAggtgac
ggtgctggttta

Tub4-R3: 5′-TATTGGGCGGTGGTAAAATTCCTGAACAAGGAAGGCATCAACTcga
tgaattcgagctcg

SAFETY NOTES

DAPI is a possible carcinogen. It may be harmful if it is inhaled, swallowed, or absorbed through the skin. It may also cause irritation. Wear gloves, facemask, and safety glasses, and do not breathe the dust.

EXPERIMENT OVERVIEW

Experiment IIA

We will familiarize ourselves with the unique sizes and shapes of yeast cells using a light microscope. Asynchronously growing haploids and diploids will be evaluated, along with cells in the process of shmooing and forming zygotes.

Experiment IIB

We will visualize yeast organelles with small fluorescent molecules, including DAPI, Mitotracker Red, rhodamine-phalloidin, calcofluor, and FM 4-64, using fluorescence microscopy.

Experiment IIC

We will use fluorescence microscopy to monitor spindle dynamics in a series of cell division cycle (*cdc*) mutants. The mutant genes in our study cause cells to arrest with bud sizes approaching those of the mother. We will use patterns of spindle and/or spindle pole body (SPB) fluorescence to define the cell cycle defect that gives rise to these similarly shaped yeast mutants. In this section, we will also fuse mCherry to γ-tubulin in a strain that bears a chromosome tagged with GFP-Lac repressor. We will use the fluorescent proteins to monitor chromosome segregation in live cells.

EXPERIMENTAL PROCEDURES

▶ Day 1

Experiment IIA

You will be provided with cultures of strain 2-1 that were fixed with 4% para-formaldehyde during either log phase growth or stationary phase. Examine these cultures using differential interference contrast (DIC) microscopy and count 100 cells of each culture, noting the numbers of unbudded, small budded, and large budded cells in the table below.

Strain	Unbudded	Small budded	Large budded	Total
2-1 log				100
2-1 stationary				100

You will also be provided with a mixture of strains 2-1 and 2-2 that were fixed while mating. Examine this culture using DIC microscopy and identify shmoos and zygotes, based on their distinctive morphology.

Draw pictures of shmoos

Draw pictures of zygotes

Compare the morphologies of haploid 2-1 and diploid 2-3 cells, looking for differences in the size and shape of the cells. Draw a picture of each cell type on the same scale.

Picture of 2-1

Picture of 2-3

Experiment IIB

In the morning, start overnight cultures of strains 2-4 and 2-5 in YPD. Shake at 30°C.

▶ *Day 2*

Experiment IIB

In the morning, dilute back strains 2-4 and 2-5 and grow back to mid-log phase (~ 1–2×10^7/mL). See Techniques and Protocols 6 (Yeast Vital Stains).

1. *Mitochondria:* Use Mitotracker Red with 1 mL of strain 2-4.

2. *Bud scars:* Use calcofluor with 1 mL of strains 2-4 and 2-5.

3. *Vacuoles:* Use FM 4-64 with 1 mL of strain 2-4. Greater specificity will be achieved with longer staining times (1–2 h).

Experiment IIC

Fusing mCherry to Tub4: TUB4 encodes γ-tubulin, a protein that nucleates microtubules at SPBs. Prepare an mCherry PCR product according to Techniques and Protocols 1 using primers Tub4-F5 and Tub4-R3 and template pFA6a-link-mCherry. Pour agarose gel and wrap in Saran wrap and refrigerate overnight. In the morning before lecture, load your gel to check for the PCR product.

▶ *Day 3*

Experiment IIB

Actin staining: You will be given strain 2-1 cells that have been fixed in paraformaldehyde according to Techniques and Protocols 7 (Actin Staining of Fixed Yeast Cells). Complete the rhodamine-phalloidin staining procedure starting at Step 9. Examine the cells using the rhodamine filter set on the fluorescence microscope.

Experiment IIC

1. *Fusing mCherry to Tub4:* Confirm the mCherry PCR product on an agarose gel. If you did not get a product, set up a second reaction as soon as possible. Start a 5 mL culture of strain 2-10 in YPD and grow overnight at 30°C for transformation.

2. *cdc mutant analysis:* In the morning, start 5-mL cultures of strains 2-6, 2-7, 2-8, and 2-9 in SC + ade to visualize spindle morphology of *cdc* mutants. Grow these *temperature-sensitive strains* overnight at 23°C in SC + ade. The extra

adenine in this media suppresses the formation of the highly fluorescent red pigment in *ade2* strains.

► Day 4
Experiment IIC

In the morning, check the density of overnight cultures. Dilute back for regrowth to mid-log phase if necessary.

1. *cdc mutant analysis:* When cultures reach mid-log phase, transfer cultures of strains 2-6 to 2-9 to a shaking, 37°C water bath. After 3 h, pellet 1 mL of each culture in a 1.5-mL Eppendorf tube and give to a TA who will fix the cells for you (see Techniques and Protocols 6, Yeast Vital Stains: DAPI staining of formaldehyde-fixed cells. The TA will omit the ethanol step for these strains). Perform this step quickly to avoid recovery of cell growth at room temperature.

2. Pellets of fixed cells will be returned to you. Resuspend the cells in a 50 µL solution of 50 ng/mL DAPI. Mount cells on agarose pads to evaluate by fluorescence microscopy (Techniques and Protocols 8, Preparation of Slides with Agarose Pads for Imaging of Live Yet Immobile Yeast). *Label the slides and think about genotypes before heading to the microscope*. Collect images for all the strains using the red channel for Tub4-mCherry, the green channel for GFP-Tub1, the blue channel for DAPI/DNA, and DIC or brightfield to record cell shape. Remember that strain 2-8 lacks GFP-Tub1. Note in your lab book the differences in morphology of the four strains.

   How do the differences correspond to what can be gleaned from SGD? _____

3. *Fusing mCherry to Tub4:* Dilute overnight culture of strain 2-10 into 50 mL of YPD (1/100) and grow to mid-log phase. Transform half of your PCR product using the high-efficiency LiOAc transformation protocol (Techniques and Protocols 1). Be sure to include a no-DNA control. Plate both transformations on YPD plates at 30°C.

► Day 5
Experiment IIC

Fusing mCherry to Tub4: Replicate your transformation plates to YPD + G418. Incubate the plates at 30°C.

► Day 7
Experiment IIC

Fusing mCherry to Tub4: Transformants should be visible on your YPD + G418 plate that had your PCR product. Search for proper integrants by fluorescence microscopy. To this end, use a sterile pipette tip to dab a small segment of the colony into 4 µL of H_2O on a glass slide. Up to three colonies can be examined per slide. Add coverslips and seal with nail polish. Repeat until one positive has been identified. It may be useful to compare the transformants to the parental strain. Restreak at least one proper integrant for single colonies on a YPD + G418 plate.

How many transformants did you screen before finding a proper integrant? _____

▶ *Day 8*

Experiment IIC

Fusing mCherry to Tub4: Inoculate the Tub4-mCherry integrant in 5 mL of SC + ade and grow overnight at 30°C.

▶ *Day 9*

Fusing mCherry to Tub4: Dilute back your cells into 5 mL of SC + ade and continue growing at 30°C for 4–8 h to get mid-log phase cells. Centrifuge 1 mL of culture, resuspend in 35 µL of SC + ade, and then place 4 µL onto a slide with an agarose pad. Seal the coverslip with nail polish. Find a region of the pad that is dense with cells, including some cells with large buds that have not yet divided. Visualize the mCherry-tagged SPBs and GFP-tagged chromosome (Lac sites at *lys2*) locus using the rhodamine and GFP filters, respectively. Try to visualize a mitosis by obtaining time-lapse images with a short z-stack for each time point.

MATERIALS

Day 1	Master plate with strains 2-4 and 2-5
	2 Culture tubes with 5 mL of YPD
	Fixed cells of strain 2-1 (mid-log and saturated), strain 2-3, and a mating mixture of strains 2-1 and 2-2
	Glass slides, coverslips, and nail polish
Day 2	2 Culture tubes with 5 mL of YPD
	3 µL of Mitotracker Red (1 mg/mL stock in DMSO; Life Technologies M7512)
	10 mL of ddH_2O
	100 µL of calcofluor (1 mg/mL stock in ddH_2O; Sigma F3543-1G)
	2 µL of FM 4-64 (1 mg/200 µL stock in ddH_2O; Molecular Probes T-3166)

Slides, coverslips, and nail polish
PCR reagents (see Techniques and Protocols 1)
   Primers: Tub4-F5, Tub4-R3
   pFA6a-link-mCherry
   Q5 DNA polymerase
   5× PCR buffer
   10 mM dNTPs
Gel electrophoresis reagents (see Techniques and Protocols 1)
   Ultrapure agarose
   1× TBE
   1-kb ladder
   Loading dye
SYBR Safe DNA stain in trough

Day 3

Fixed culture of strain 2-1 using the rhodamine-phalloidin protocol
20 µL of rhodamine-phalloidin (6.6 µM stock in MeOH; Molecular
   Probes R-415; aliquot and store at −20°C)
   10 mL of PBS
   5 µL of DAPI (1 mg/mL stock in ddH$_2$O; Sigma D9542; aliquot and
   store at −20°C)
Gel-electrophoresis reagents (see Techniques and Protocols 1)
   1-kb Plus Ladder
   Gel-loading buffer
   1× TBE
   SYBR Safe DNA stain in trough
Master plate with strains 2-6 to 2-10
1 Culture tube with 5 mL of YPD
4 Culture tubes with 5 mL of SC + ade

Day 4

4 Culture tubes with 5 mL of SC + ade
5 µL of DAPI (1 mg/mL stock in ddH$_2$O; Sigma D9542; aliquot and
   store at −20°C)
5 mL of PBS
Depression well microscope slides (Fischer 50-949-458)
Agarose gel aliquots (see Techniques and Protocols 8)
Coverslips and nail polish
50 mL of YPD in a 250-mL Erlenmeyer flask
Transformation reagents (see Techniques and Protocols 2)
   LiOAc in TE
   Salmon sperm DNA
   PEG solution

DMSO
TE
Sterile glass beads
2 YPD plates

Day 5   2 YPD + G418 plates

Day 7   1 YPD + G418 plate

Day 8   Coverslips
H_2O
Glass slides
1 Culture tube with 5 mL of H_2O SC + ade

Day 9   1 Culture tube with 5 mL of SC + ade
Depression well microscope slides (Fischer 50-949-458)
Agarose gel aliquots (see Techniques and Protocol 8)
Coverslips and nail polish

REFERENCES

Hartwell LH, Mortimer RK, Culotti J, Culotti M. 1973. Genetic control of the cell division cycle in yeast: V. Genetic analysis of *cdc* mutants. *Genetics* **74:** 267–286.

Huh WK, Falvo JV, Gerke LC, Carroll AS, Howson RW, Weissman JS, O'Shea EK. 2003. Global analysis of protein localization in budding yeast. *Nature* **425:** 686–691.

Straight AF, Belmond AS, Robinett CC, Murray AW. 1996. GFP tagging of budding yeast chromosomes reveals that protein-protein interactions can mediate sister chromatid cohesion. *Curr Biol* **6:** 1599–1608.

Vida TA, Emr SD. 1995. A new vital stain for visualizing vacuolar membrane dynamics and endocytosis in yeast. *J Cell Biol* **128:** 779–792.

Manipulating Mating-Type and Epigenetic Transcriptional Silencing

The ability of *Saccharomyces cerevisiae* to reproduce sexually affords the opportunity to mix and match genes via genetic crosses. Moreover, the ability of yeast to reproduce in haploid form greatly facilitates genetic analyses. In this section, we learn about the genes that determine mating-type, how mating-type can be switched, and the epigenetic mechanism that maintains the auxiliary mating-type loci in a transcriptionally silenced state. To illustrate one utility of mating, we perform a two-hybrid assay for protein–protein interactions by mating haploids that contain bait and prey vectors.

MATING-TYPE

S. cerevisiae has both asexual and sexual reproduction cycles. During vegetative growth, both haploid and diploid cells reproduce asexually by budding. Haploid cells are also able to reproduce sexually by mating with each other to generate diploid cells. Upon conditions of stress, such as nitrogen starvation, diploid cells undergo meiosis to regenerate haploid cells.

Mating of haploid yeast is governed by a cell's mating-type, which is determined by expression of mating-type genes at the *MAT* locus (Fig. 1). Cells bearing the *MAT***a** allele are designated **a** cells; cells bearing the *MAT*α allele are designated α cells. Cells of the same mating-type are unable to mate with each other. Mating of **a** cells with α cells gives rise to diploids that contain both *MAT***a** and *MAT*α alleles. The resulting diploids are designated **a**/α.

The stability of mating-type classifies strains into two groups: heterothallic and homothallic. Heterothallic strains have a stable mating-type, whereas homothallic haploid strains are capable of changing the mating-type. Therefore, a colony derived from a homothallic cell will contain both **a** and α cells as well as diploid cells that arise from matings between **a** and α cells. Most laboratory strains are heterothallic due to a mutation in a single gene designated *HO*. In this set of experiments, we

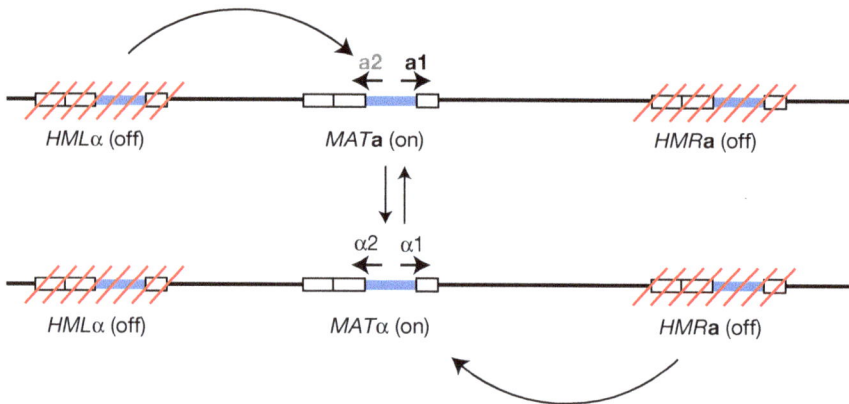

FIGURE 1. The active mating-type locus *MAT* can undergo switching using the sequences encoded at the silenced mating loci *HML* and *HMR*. Silencing is indicated by diagonal lines. (Adapted from Haber 1998, with permission, copyright Annual Reviews, http://www.annualreviews.org.)

express *HO* from the inducible *GAL1* promoter to switch a set of strains from one mating-type to another.

The ability of cells to switch mating-types is dependent on two other loci: *HML* and *HMR*, located on the left and right arm of chromosome III, respectively (Fig. 1). These loci contain unexpressed but complete copies of the mating-type genes. In general, *HML* has the α cell type information and *HMR* has the **a** cell type information. *HO* encodes a site-specific endonuclease that cuts only within the *MAT* locus. The resulting double-strand break stimulates recombination between the *MAT* locus and either *HML* or *HMR*. The *MAT* allele (**a** or α) determines whether *HML* or *HMR* is used as a donor. As this recombination event is not reciprocal, the information from the silent locus remains intact while the information at the *MAT* locus changes. These observations led to the cassette model for mating-type switching (Fig. 1). The α cassette contains the *α1* and *α2* genes whereas the **a** cassette contains the **a***1* and **a***2* genes. These genes (with the exception of **a***2*, which has yet to be assigned a function) encode transcriptional regulators that control expression of **a**-specific, α-specific, haploid-specific, and diploid-specific genes (see example in Fig. 2).

HML AND *HMR* SILENCING

Analysis of mating-type in budding yeast has been used to identify transcription activators and repressors, recombination proteins, genes involved in chromatin structure, budding, polarity, and pseudohyphal growth factors and proteins involved in cell cycle control and the secretory pathway (reviewed in Madhani 2007). This is because the genes expressed at *MAT* (and repressed at *HML* and *HMR*) initiate pathways that ultimately govern all of these cellular processes.

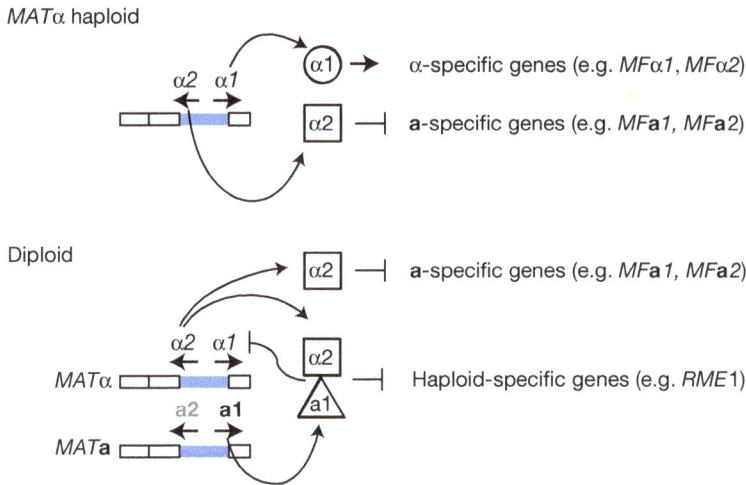

FIGURE 2. Control of expression of cell-type-specific gene sets by regulatory proteins encoded by *MAT*. In α cells, the α1 protein activates α-specific genes. The α2 protein represses transcription of **a**-specific genes. In diploids, the α2 and **a1** proteins repress haploid- and **a**-specific genes, as well as α1. In the absence of α1, α-specific genes are not activated. (Adapted from Herskowitz 1989, with permission from Macmillan Publishers Ltd.)

A heterochromatin-like structure termed silent chromatin is required for transcriptional silencing of *HML* and *HMR*. Silent chromatin consists of nucleosomes and four *silent-information regulators* (Sir1, Sir2, Sir3, and Sir4) that assemble in two operationally defined steps. In the first step, Sir1 binds specific sites at *HML* and *HMR* and recruits the other Sir proteins to nucleate silent chromatin assembly. In the second step, Sir2–4 spread across the locus through the action of Sir2, a histone deacetylase, and Sir3–4, which bind deacetylated histone tails. Silent chromatin also assembles in subtelomeric domains.

VARIEGATED EXPRESSION AND EPIGENETIC INHERITANCE

Mutants that cause inefficient nucleation of silencing yield mixed populations of cells with variegated repression of the *HM* loci. Some cells are silent, whereas other genetically identical cells are not. Moreover, these two opposing expression states are inherited faithfully in an epigenetic fashion. The simplicity of silencing assays and the power of yeast as an experimental system has made yeast mating-type silencing a premier model for chromatin-based epigenetic phenomena.

ASSAYS TO MONITOR MATING-TYPE SILENCING

Diploids do not mate because expression of the **a** and α mating-type genes at *MAT* blocks expression of haploid-specific genes necessary for mating (Fig. 2).

Derepression of the *HM* loci in haploids creates a pseudodiploid state in which both **a** and α mating-type genes are expressed. Thus, silencing mutants also fail to mate, a phenotypic consequence that has been useful in identifying modifiers of the silencing pathway.

Unbudded haploid yeast cells arrest growth with a distinct shmoo morphology when exposed to mating pheromone from cells of the opposite mating-type. Thus, pheromone treatment can be used to determine whether the *HM* loci of haploid cells are silenced. The pheromone produced by *MATα* haploids (α factor) is soluble and commercially available. Because binding and cell cycle arrest are reversible, the α factor oligopeptide is commonly used to create cultures of synchronously growing *MATa* cells.

Silencing of *HML* and *HMR* can also be measured by inserting reporter genes at one or both of these loci. *URA3* is commonly used. In this case, cells that silence the reporter gene at *HM* loci grow well on plates containing 5-FOA (5-fluoro-orotic acid) which is toxic to cells expressing *URA3*. In contrast, cell growth on plates lacking uracil is hindered. Thus, mutants that lack silencing grow well on SC-ura plates but die on 5-FOA. This powerful selection has been used in genetic screens for mutants that directly affect silencing, as well as mutants that incur pleiotropic chromatin defects.

In the experiments described here, a chimeric *URA3-GFP-NLS* reporter at *HML* will be used to investigate silencing, silencing mutants, and variegated expression (Fig. 3). This reporter construct permits evaluation of *HML* in two distinct ways: (1) semiquantitative plating assays will be used to measure expression of *URA3* in bulk populations and (2) fluorescence microscopy will be used to measure expression of green fluorescent protein (GFP) in single cells. The nuclear localization signal

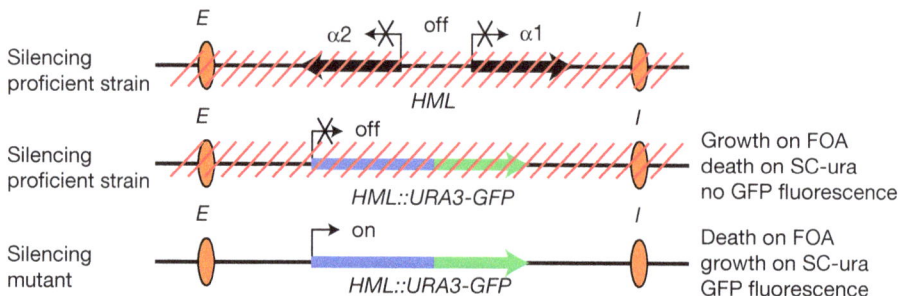

FIGURE 3. Reporter system to monitor silencing at *HML*. The α*1* and α*2* mating-type genes at *HML* were replaced with a *URA3-GFP-NLS* expression cassette. The gene is silenced in wild-type cells due to the Sir proteins and *cis*-acting nucleation sites, *E* and *I*. Loss of silencing of *URA3-GFP-NLS* causes death when cells are exposed to 5-FOA but permits survival of cells with green nuclear fluorescence on media lacking uracil (Xu et al. 2006; Kitada et al. 2012).

(NLS) of the construct concentrates GFP in nuclei of nonsilent cells. A second marker at *HMR* in this strain, *mCherry-NLS*, permits comparison of silencing at the two *HM* loci in single cells.

PRACTICAL APPLICATIONS OF MATING YEAST

In addition to the remarkable utility that mating offers in crossing different genetic lineages, the ability to mate strains gives the experimenter an easy way to transfer reporter plasmids. High-throughput two-hybrid screens, where bait and prey vectors were paired by mating, serve as illustrative examples. A genome-wide interaction study by Uetz et al. used two-hybrid reporter strains that contained three different reporter genes fused to promoters driven by the Gal4 transcriptional activator (James et al. 1996; Uetz et al. 2000). Bait proteins were fused to the Gal4 binding domain in strains of one mating-type, whereas prey proteins were fused to the Gal4 activation domain in strains of the opposite mating-type. After mating, interactions between bait and prey proteins reconstituted Gal4, resulting in activation of the three reporter gene fusions (*GAL1p-HIS3*, *GAL2p-ADE2*, *GAL7p-lacZ*). In the experiments described here, we will mate two-hybrid reporter strains to validate the interaction between Rtt107 and Slx4, two proteins involved in DNA repair.

STRAINS

3-1 HS2	*MAT**a** hom3*
3-2 HS3	*MATα hom3*
3-3 MRG4361	*MATα hmr::URA3p-mCherry-NLS hml::URA3-GFP-NLS ade2-1*
3-4 MRG4362	*MATα hmr::URA3p-mCherry-NLS hml::URA3-GFP-NLS ade2-1 Δsir1*
3-5 MRG4363	*MATα hmr::URA3p-mCherry-NLS hml::URA3-GFP-NLS ade2-1 Δsir2*
3-6x W303	*MAT**?** with pGAL-HO(URA3)*
3-6y W303	*MAT**?** with pGAL-HO(URA3)*
3-8 PJ69-4A	*MAT**a** trp1-901 leu2-3 ura3-52 his3-200 gal4 gal80 GAL2p-ADE2 LYS2::GAL1p-HIS met2::GAL7p-lacZ* with pBAIT (LEU2)
3-9 PJ69-4A	*MAT**a** trp1-901 leu2-3 ura3-52 his3-200 gal4 gal80 GAL2p-ADE2 LYS2::GAL1p-HIS met2::GAL7p-lacZ* with pBAIT-Rtt107 (LEU2)

| 3-10 PJ69-4α | *MATα trp1-901 leu2-3 ura3-52 his3-200 gal4 gal80 GAL2p-ADE2 LYS2::GAL1p-HIS met2::GAL7p-lacZ* with pPREY (TRP1) |
| 3-11 PJ69-4α | *MATα trp1-901 leu2-3 ura3-52 his3-200 gal4 gal80 GAL2p-ADE2 LYS2::GAL1p-HIS met2::GAL7p-lacZ* with pPREY-Slx4 (TRP1) |

Note: Most laboratory strains are heterothallic due to an ho *mutation, but that mutation is rarely listed in the strain genotypes. Strains 3-1 and 3-2 are useful for testing mating types because they contain a rare auxotrophic mutation. Otherwise, the genetic background is unknown. Therefore, these strains should never be used for genetic crosses other than mating-type testing.*

PLASMIDS

pGAL-HO-(URA3)

EXPERIMENT OVERVIEW

Experiment IIIA

We will express *HO* ectopically to initiate mating-type switching events and evaluate our success through mating assays. This procedure generates isogenic strains that only differ by mating type. The experiment illustrates the utility of the *GAL1* promoter in driving rapidly inducible expression of a heterologous gene.

Experiments IIIB and IIIC

We will monitor reporter genes to evaluate the impact of different *SIR* genes on silencing the mating-type cassettes. We will use semiquantitative plating assays and microscopy to illustrate the variegated expression states and the difference between phenotypic properties of single cells versus bulk populations.

Experiment IIID

We will demonstrate the utility of mating and the power of the two-hybrid scheme by pairing various bait and prey vectors. Plating assays will be used as readouts of genetic interactions.

EXPERIMENTAL PROCEDURES

▶ *Day 1*

Experiment IIIA

Streak out pGAL1-HO-(URA3)-bearing haploid strain 3-6(x or y) for single colonies on SC-ura. The mating type of your haploid strain has been purposely concealed. Incubate at 30°C. Expression of *HO* in the haploid strain will switch the mating type.

Experiments IIIB and IIIC

Streak out strains 3-3, 3-4, and 3-5 with reporter genes at the *HM* loci on YPD. Incubate at 30°C.

Experiment IIID

Streak out two-hybrid strains 3-8 and 3-9 on SC-leu and strains 3-10 and 3-11 on SC-trp. Incubate at 30°C. pBAIT is an empty bait vector expressing only the Gal4-binding domain, whereas pPREY is an empty prey vector expressing only the Gal4 activation domain. pBAIT-Rtt107 and pPREY-Slx4 are related vectors that contain gene fusions to the two interacting factors. Save all of these master plates until the experiments are complete.

▶ *Day 3*

Experiment IIIA

In the morning, inoculate strain 3-6 into 5 mL of SC-ura + 2% dextrose and place on a shaker at 30°C. In the late afternoon, use these precultures to inoculate 5 mL of SC-ura + 2% raffinose at a 1/100 dilution. When cells are grown on dextrose, the *GAL1* promoter is repressed. Growth on raffinose removes this repression, which is a prerequisite for rapid induction (1000×) when galactose is added. Low-level expression in raffinose sometimes yields "leaky" switching events before adding galactose. We will watch for these with a preinduction time point.

Experiment IIID

To initiate mating, patch the two-hybrid strains (3-8, 3-9, 3-10, and 3-11) onto a YPD plate either alone or in the following pairwise combinations: 3-8 × 3-10, 3-8 × 3-11, 3-9 × 3-10, and 3-9 × 3-11.

▶ *Day 4*

Experiment IIIA

In the morning, dilute the overnight culture of strain 3-6 to 0.1 OD in 5 mL of SC-ura + raffinose media and grow for about two more doublings. When the cells reach mid-log phase (~0.4–0.5 OD), plate ~200 cells on a YPD plate assuming ~2×10^7 cells per OD. As a rough guide, dilute your cells ~1:100 and then use this to make a second 100-fold dilution. Plate 200 µL of this final dilution. Incubate plates at 30°C.

To the remainder of the cultures of strain 3-6, initiate *HO* expression by adding galactose to a final concentration of 2% and shake in an incubator at 30°C. After 90 min, dilute cells as before and plate 200 cells on a YPD plate. The dextrose in these plates will repress further *HO* expression from the *GAL1* promoter. Incubate plates at 30°C.

Experiment IIIB

Inoculate 5 mL of YPD with strains 3-3, 3-4, and 3-5 for overnight growth at 30°C.

Experiment IIID

Select for mated diploids from the series of two-hybrid mating reactions by replica plating to an SC-leu,trp plate.

▶ *Day 5*

Experiment IIIB

Cultures typically reach about equivalent cell densities after overnight growth to saturation in liquid YPD. Assess cell density by measuring OD_{600} of diluted cultures. Adjust densities to be equivalent if necessary. Centrifuge 1 mL of cells from each culture of strains 3-3, 3-4, and 3-5. Resuspend in 1 mL of sterile H_2O.

Silencing of the *URA3-GFP-NLS* reporter gene at *HML* will be monitored with a semiquantitative spotting assay for *URA3* expression. Use a 96-well microtiter plate (BD Falcon 35177) to set up dilutions (Fig. 4). Add 200 µL of sterile H_2O to columns 2 through 6 using a multichannel pipette. Transfer 220 µL of strain 3-3 to the first well of row A. Repeat this procedure with the remaining strains 3-4 and 3-5 in rows B and C. Create 10-fold serial dilutions along the rows by transferring 20 µL of culture from column 1 to column 2. Pipette up and down to mix and repeat the process through to the last column, from which 20 µL should be discarded.

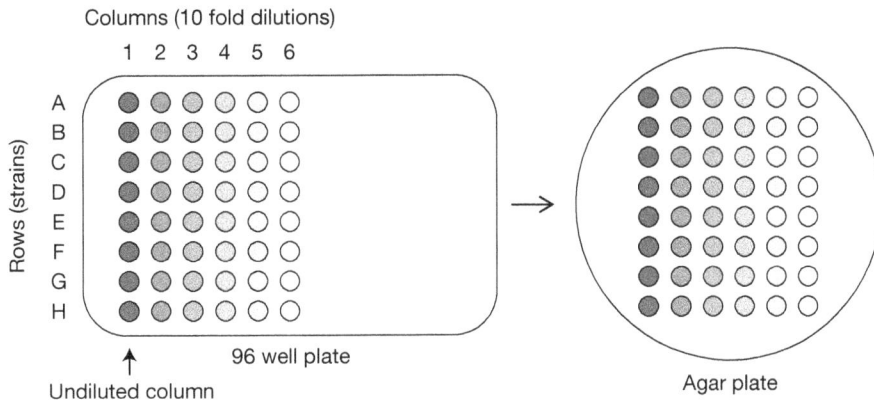

FIGURE 4. Loading 96-well microtiter plate with 10-fold dilutions (columns) for semiquantitative analysis of phenotype of strains (rows) and transfer to agar Petri dish.

Sterilize a 48-pin frogging device in ethanol, flame, and then cool the device in a Petri dish containing sterile H_2O. Spot the serial dilutions by frogging onto the following plates: (1) SC to measure loading; (2) SC-ura to measure *URA3* expression by direct selection; (3) SC + FOA to measure *URA3* expression by counter-selection; (4) SC + FOA + nicotinamide to test whether *URA3* repression is dependent on Sir2, which is inhibited by nicotinamide. Allow the plates to dry before moving them to a 30°C incubator.

▶ Day 6

Experiment IIIC

Silencing of the *URA3-GFP-NLS* reporter gene at *HML* and the *mCherry-NLS* cassette at *HMR* will be monitored by fluorescence of individual cells using microscopy. Start 5-mL cultures of strain 3-3, 3-4, and 3-5 from single colonies in SC + ade media and shake overnight at 30°C.

▶ Day 7

Experiment IIIA

We will test the mating type of cells from the *GAL1p-HO* induction by mating to known standards, strain 3-1 (*MAT**a***) and strain 3-2 (*MAT*α). To this end, use a toothpick to create two long parallel streaks of cells on a YPDA plate, one of strain 3-1 and the other of strain 3-2. Cross-streak each of the standards with five colonies from your 0 time point. This will determine the initial mating type of your unknown. Create at least five additional cross-streaks with cells from your 90-min galactose induction. This will determine whether the isolates switched mating type. *In each case, do not use the same toothpick to cross both mating standards.*

Incubate overnight at 30°C.

Experiment IIIB

Evaluate relative silencing of *URA3* in each of the three plated strains. In your lab book, record the following observations of the strains:

- Do the *sir1* and *sir2* mutants disrupt silencing equally? What can we conclude about *sir1*?

- Do SC-ura and SC + FOA plates report silencing equally? If so, how can the difference be explained?

- Is silencing compromised by nicotinamide?

Experiment IIIC

Dilute back overnight cultures of strains 3-3, 3-4, and 3-5 to 0.1 OD in 5 mL of SC + ade media and grow for about two more doublings to ~0.4 OD. Make at least three slides with agarose pads according to Techniques and Protocols 8 (Preparation of Slides with Agarose Plugs for Imaging of Live Yet Immobile Yeast). Centrifuge 1 mL of each cell culture and resuspend in 50 µL of SC + ade media. Place 3 µL of each culture on each agarose pad, add a coverslip, and seal with nail polish. Use bright-field illumination and fluorescence microscopy with the GFP and rhodamine filters to visualize the subset of cells with a transcriptionally active *HML* and *HMR* loci.

In your lab book, record the following observations of the strains:

- Do any of the nuclei fluoresce green? Red?

- For strains with red or green fluorescence, do all of the cells of the strain fluoresce?

- For strains with red or green fluorescence, is the intensity of fluorescence roughly equivalent (qualitatively) for each fluorescing cell?

- How do the GFP fluorescence results correspond with the *URA3* plating assays?

Experiment IIID

Use the SC-leu,trp replica-plate as a source of mated cells and inoculate 5 mL of SC-leu,trp with the following crosses: 3-8 × 3-10, 3-8 × 3-11, 3-9 × 3-10, and 3-9 × 3-11.

▶ *Day 8*

Experiment IIIA

Replica-plate your mating tests to an SD plate. Incubate at 30°C.

Experiment IIID

Typically, more than one reporter gene is used in two-hybrid testing to eliminate false positives that are specific to the reporter pathway. In the interest of time, score the Rtt107/Slx4 interaction with only the *GAL1p-HIS3* reporter construct by semi-quantitative spotting. Normalize the cell densities of each of the overnight cultures, centrifuge 1 mL of each, and resuspend in 1 mL of H_2O. Pipette 200 µL of undiluted culture into the first column of a 96-well microtiter plate. Use a multichannel pipette to create 10× serial dilutions in the adjacent five columns. The dilution series will be frogged to (1) SC to measure loading, (2) SC-his to test for interactions, and (3) SC-his,+ 1 mм 3-AT to suppress background. 3-AT competitively inhibits the *HIS3* gene product, thereby raising the threshold level of *HIS3* expression sufficient for growth in the absence of histidine. Note that the only selection for plasmids in these plates is based on the interaction between bait and prey. Although it is possible to select for plasmids and an interaction with SC-trp,-leu,-his, the triple drop-out media imposes an unnecessary additional burden on growth.

▶ *Day 9*

Experiment IIIA

Score the mating type of isolates from your *GAL-HO* time course. The results from your time = 0 plate will tell you the original mating type of strain 3-6. What might be the explanation for strains that do not mate to either of the test strains 3-1 or 3-2?

Strain, time	# mated to 3-1	# mated to 3-2	# mated to neither	Starting mating type	Final mating type
3-6, $t = 0$					N/A
3-6, $t = 90$				N/A	

▶ *Day 10*

Experiment IIID

Remove the SC plates from the incubator and store at 4°C. These loading controls grow faster than the selection plates and must be pulled early to prevent overgrowth.

▶ *Day 12*

Experiment IIID

Evaluate the two-hybrid reporter plates for interactions between Rtt107 and Slx4 versus the two proteins with control vectors (pBAIT and pPREY alone). Did 3-AT aid in the characterization of this interaction? What value would have been added by using the *GAL2p-ADE2* and *GAL7p-lacZ* reporters?

MATERIALS

Day 1   1 SC-ura plate
1 SC-leu plate
1 SC-trp plate
1 YPDA plate
Strains: 3-3 through 3-11

Day 3   1 Culture tube with 5 mL of SC-ura
1 Culture tube with 5 mL of SC-ura + raffinose (no dextrose)
1 YPD plate

Day 4   1 Culture tube with 5 mL of SC-ura + raffinose
2 YPD plates
20% galactose
3 Culture tubes with 5 mL of YPD
1 SC-leu,trp plate

Day 5   Sterile 96-well microtiter plate (BD Falcon 35177)
48-pin frogging device
1 SC plate
1 SC-ura plate
1 SC + FOA plate
1 SC + FOA +10 mM nicotinamide plate

Day 6   3 Culture tubes with 5 mL of SC + ade

Day 7   1 YPD plate
Strains 3-1 and 3-2
3 Culture tubes with 5 mL of SC + ade
Depression-well microscope slides (Fisher 50-949-458)

Agarose gel aliquots (see Techniques and Protocols 8, Preparation of Slides with Agarose Pads for Imaging of Live Yet Immobile Yeast)

4 Culture tubes with 5 mL of SC-leu,trp

Day 8   1 SD plate

1 SC plate

1 SC-his plate

1 SC-his,+1 mM 3AT

REFERENCES

Haber JE. 1998. Mating-type gene switching in *Saccharomyces cerevisiae*. *Annu Rev Genet* **32:** 561–599.

Herskowitz I. 1989. A regulatory hierarchy for cell specialization in yeast. *Nature* **342:** 749–757.

James P, Halladay J, Craig EA. 1996. Genomic libraries and a host strain designed for highly efficient two-hybrid selection in yeast. *Genetics* **144:** 1425–1436.

Kitada T, Kuryan BG, Tran NN, Song C, Xue Y, Carey M, Grunstein M. 2012. Mechanism for epigenetic variegation of gene expression at yeast telomeric heterochromatin. *Genes Dev* **26:** 2443–2455.

Madhani HD. 2007. *From a to alpha: Yeast as a model for cellular differentiation*. Cold Spring Harbor Laboratory Press, New York.

Uetz P, Giot L, Cagney G, Mansfield TA, Judson RS, Knight JR, Lockshon D, Narayan V, Srinivasan M, Pochart P, et al. 2000. A comprehensive analysis of protein-protein interactions in *Saccharomyces cerevisiae*. *Nature* **403:** 623–627.

Xu EY, Zawadzki KA, Broach JR. 2006. Single-cell observations reveal intermediate transcriptional silencing states. *Mol Cell* **23:** 219–229.

Mating, Meiosis, and Tetrad Dissection

MEIOSIS

As part of the life cycle of *Saccharomyces cerevisiae*, *MAT***a**/*MAT*α diploids undergo meiosis to form haploid spores (Fig. 1). The process, termed sporulation, occurs when any one of a number of essential elements becomes growth-rate-limiting (carbon source, for example). Meiosis initiates with the replication of each chromosome. Homologous chromosomes then pair, undergo recombination, and line up on the spindle apparatus. Once aligned, the homologs separate and go to an opposite pole in the first meiotic division (reductional division). In the second division, each chromosome associates with the spindle apparatus and then each chromatid goes to an opposite pole. This process, like meiosis in higher eukaryotes, generates four meiotic products (ascospores) that contain a 1C (haploid) DNA content. The four ascospores derived from a single meiotic event are contained in an ascus or sac. This sac protects the four spores from perishing during unfavorable conditions. Unlike gametogenesis in many higher eukaryotic organisms, these four haploid meiotic products (gametes) can be collected as a unit (the ascus), and all four meiotic products can be grown vegetatively as haploid cells.

TETRAD DISSECTION

The ability to grow all four haploid products of yeast meiotic events facilitates genetic analyses with this organism. The ascal wall can be partially digested by treatment with an enzyme (such as glusulase or zymolyase), releasing the four meiotic products. We can view these products and even manipulate them because they tend to be sticky and stay together as a tetrad (Fig. 2A).

After digestion of the ascus wall, the tetrahedral shape of the tetrad relaxes to a diamond shape where all four spores are visible (Fig. 2B). Tetrads placed on a YPD plate can be viewed under a specially adapted microscope. This dissecting microscope includes a micromanipulator that holds a needle made of thinly drawn-out glass or optical fiber that can be visualized under the microscope. This needle can

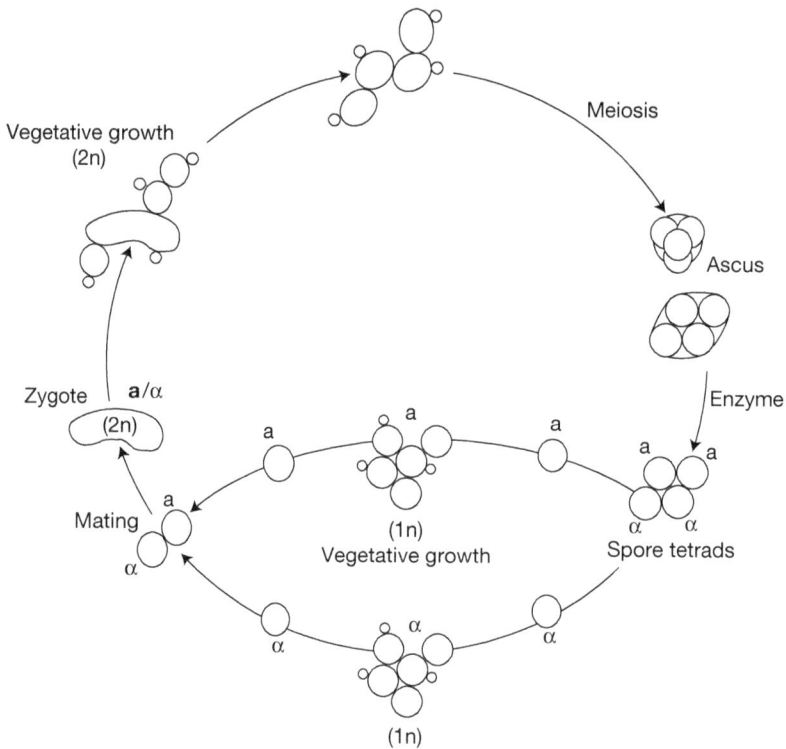

FIGURE 1. Life cycle of budding yeast.

be moved in an *x, y,* and *z* coordinate system. The thin tip of the needle can touch the agar surface and pick up a tetrad. It can then be used to separate the four meiotic products equidistant from each other and in a contiguous array. Each spore will then germinate on the YPD agar plate and grow vegetatively into a colony. Each colony represents a pure population of haploid yeast cells that are genetically identical to the single haploid meiotic product originally placed down at that position on the agar surface.

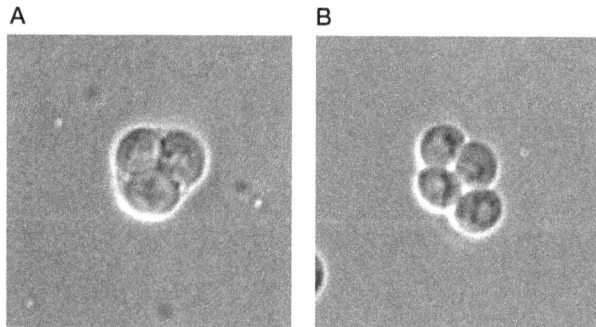

FIGURE 2. (*A*) DIC image of a tetrad. (*B*) DIC image of a tetrad after zymolyase digestion.

TETRAD ANALYSIS

Colonies can be replica plated to other types of media to analyze their growth requirements. If the original diploid strain contained an auxotrophic mutation, one can then analyze the genetic nature of the mutation that confers the auxotrophy and study its relationship to mutations in other genes. Although meiotic mapping is rarely done today to identify the location of a mutation, tetrad analysis is still a mainstay of any yeast genetics lab because it is useful to determine linkage, to demonstrate that a phenotype is caused by a defect in a single nuclear gene, and to construct strains for genetic, cytological, and biochemical analysis.

Three types of linkage relationships can be established by tetrad analysis:

1. Are the two genes genetically linked and, thus, on the same chromosome?

2. Are the two genes unlinked to each other and, thus, either far apart from each other on the same chromosome or located on different chromosomes?

3. Are the two genes unlinked to each other and located on different chromosomes but jointly linked to the centromeres of their respective chromosomes?

Analysis of meiotic progeny by random sporulation (as carried out in Experiment VII) will only provide information on the linkage of two genes to each other. It cannot be used to determine if two unlinked genes are linked to their respective centromeres. Although the isolation of four spores from an ascus is a skill only acquired with considerable practice, tetrad dissection and analysis are important tools for any yeast geneticist.

TYPES OF TETRADS

The best way to understand meiosis and tetrad analysis is to consider the segregation of markers from a cross of two haploids, $AB \times ab$ (A and B are two different genes and uppercase vs. lowercase reflect distinguishable alleles). Three classes of tetrads are observed: parental ditype (PD), nonparental ditype (NPD), and tetratype (T) (Fig. 3). The ratio of the three types of tetrads that is observed for a pair of markers is a function of the relative map positions of the markers. Thus, analysis of the frequency of PD:NPD:T tetrads can be used to deduce map distances and linkage between two genes with each other and with their centromeres.

Case 1: A and B Are on the Same Chromosome and Tightly Linked

As shown in Figure 3, when there are no crossovers between the A locus and the B locus, a PD tetrad is formed. This occurs when the distance between two genes is sufficiently small such that no recombination occurs. If we dissect tetrads and there

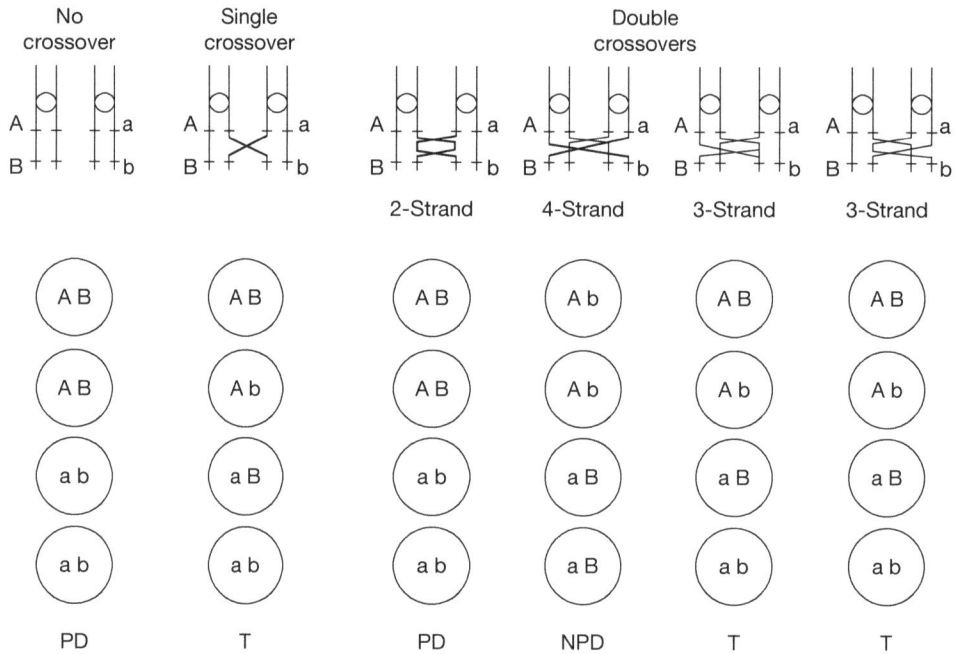

FIGURE 3. Origin of PD, NPD, and T tetrads.

are only two types of haploid meiotic products that are identical to the parental types, this tells us the genes are on the same chromosome and tightly linked.

Case 2: A and B Are on the Same Chromosome and Less Tightly Linked

If the *B* gene is further away from *A* on the same chromosome, then the chances of recombination between *A* and *B* increases. The first type of recombination event that will occur is a single crossover, which will result in the formation of a T tetrad. There are four possible types of double crossovers involving two, three, or four strands (chromatids). Two-strand double crossovers yield a PD, the two types of three-strand double crossovers both yield Ts, and a four-strand double crossover yields an NPD (Fig. 3). Meioses in which more than two crossovers occur between *A* and *B* can result in any of the three types of tetrads, depending on which strands are involved. The probability that a crossover will occur between two markers is approximately proportional to the physical distance between them.

Therefore, if *A* and *B* are linked (but not very tightly) and not completely free to recombine, and we analyze many tetrads from independent meiotic events, we will observe some PD, NPD, and T tetrads. For two genes on the same chromosome that are genetically linked, there will always be more PDs than

NPDs. To determine if the number of PD versus NPD tetrads is statistically significant, use the Chi^2 test.

Case 3: A and B Are Unlinked Genes

If A and B are very far away from each other on the same chromosome, we would expect to see a ratio of PD:NPD:T of 1:1:4. The same types of tetrads and same ratios are also seen if two genes are on different chromosomes and at least one of those genes is free to recombine with its centromere. Thus, if a ratio of 1:1:4 of PD:NPD:T is seen in your tetrads, all you can conclude is that the genes are unlinked. You cannot say if they are on the same chromosome or different chromosomes.

Case 4: A and B Are Centromere-Linked Genes

Tetrad analysis (and only tetrad analysis) can be used to detect and analyze the relationship of a pair of genes with their centromeres. When discussing centromere linkage, it is important to remember that the two genes are NOT linked to each other. Instead, A and B are on different chromosomes. As discussed in Case 3, if either A or B is unlinked to its centromere, then we would detect 1 PD:1 NPD:4 T. However, if A is very close to its centromere and B is also close, then PD and NPD tetrads will be recovered in approximately equal numbers because the two genes are unlinked (they are on different chromosomes). NPDs do not arise because of double crossovers when two genes are on different chromosomes. Rather, NPDs arise because of the alignment of the two independent chromosomes on the meiotic spindle in the first meiotic division. That alignment gives rise to PDs and NPDs with equal probability. Few tetratype asci will form because a crossover between either A or B and its centromere must occur. For centromere-linked genes, the T class is a measure of the sum of the genetic distances between each gene and its respective centromere. In classical nomenclature, the T class of tetrads forms through a process called second-division segregation, which is described in Figure 4.

TETRAD ANALYSIS AT A GLANCE

Below is a summary of the types of ratios of tetrads we should expect to see for gene pairs located at different positions within the genome. Although we can often simply look up the position of a gene in the reference genome in *Saccharomyces* Genome Database (SGD) understanding the basics of tetrad analysis is helpful in generating strains and sorting through complex phenotypes such as synthetic lethality and extragenic suppression.

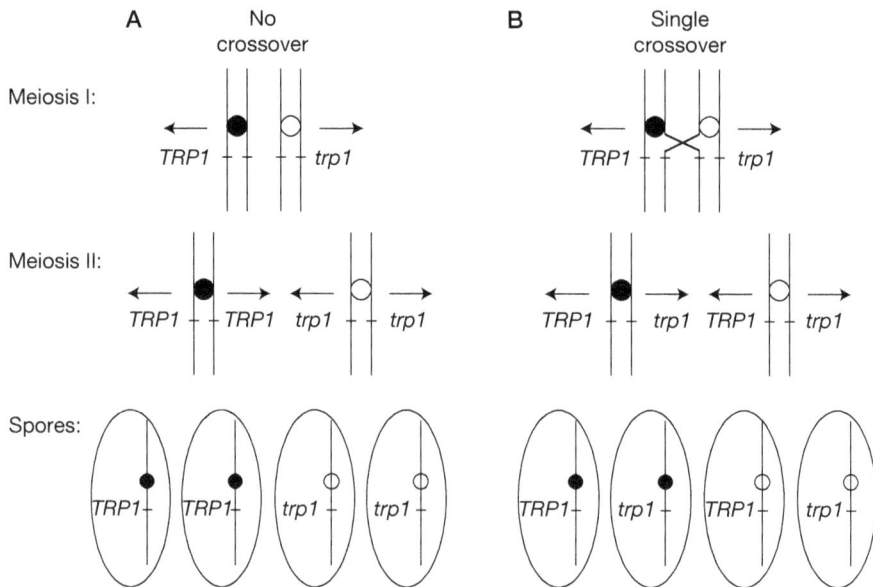

FIGURE 4. First- and second-division segregation. A gene located close to the centromere, such as *TRP1* on chromosome IV, exhibits few crossovers between the *TRP1* locus and *CEN4*, and thus, *TRP1* and *trp1* alleles segregate away from each other in the first meiotic division (*A*). This is called first-division segregation. In a fraction of meiosis, there will be a crossover between the *TRP1* locus and *CEN4* as shown in *B*. This results in homologs that carry one allele of *TRP1* and one allele of *trp1*. In this situation, the *TRP1* and *trp1* alleles do not migrate away from each other until meiosis II. This is called second-division segregation.

Genes on separate chromosomes	PD	:	NPD	:	T
1. Both inseparable from their centromeres	1		1		0
2. One very close to its centromere, the other near but separable	1		1		<4
3. One very close to its centromere, the other far	1		1		4
4. Both very far from their centromeres	1		1		4

Genes on the same chromosome					
5. Genes are very close	Mainly PD		0		a few
6. Closely linked	PD	>	NPD		
7. Far from each other	1		1		4

TETRAD DISSECTION IN PRACTICE

In this experiment, you will analyze tetrads to determine gene and centromere linkage. You will also learn how to determine the mating type of haploid progeny using complementation of a rare auxotrophy (*his1*) with tester strains. This plate-based assay is known as the mating-type test (also see Experiment IIIA).

You will also learn how to construct strains using meiosis and tetrad dissection. Often, it is necessary to build strains containing the same auxotrophic marker used at different loci, and in this experiment, we will construct a strain containing *SPC42-mCherry* and *URA3* at the *SPC42* locus and *NUP49-mTurquoise* and *URA3* at the *NUP49* locus. In yeast nomenclature, this is written as *SPC42-mCherry::URA3* and *NUP49-mTurquoise::URA3*. The parental strains also contain additional auxotrophic markers that will allow us to select for diploids and determine positional information about a temperature-sensitive mutation that is present in one of these strains. Following selection of diploids, it is your job to sporulate, dissect, and analyze your resulting tetrads to find haploids that contain both *SPC42-mCherry::URA3* and *NUP49-mTurquoise::URA3* in the same strain.

Finally, you will learn how to isolate diploids from a mating mixture by two methods: by using auxotrophic markers and by pulling zygotes.

STRAINS

4-4	*MAT**a** his1*
4-5	*MATα his1*
4-6	*MAT**a** cdc15-2 SPC42-mCherry::URA3 TRP1 ade1 ura3-1 leu2-3,112 his3-11,15*
4-7	*MATα NUP49-mTurquoise::URA3 trp1-1 ura3-1 leu2-3,112 his3-11,15 ADE1*
4-8	4-6 × 4-7
BY4741	*MAT**a** ura3Δ0 leu2Δ0 his3Δ1 met15Δ0*
BY4742	*MATα ura3Δ0 leu2Δ0 his3Δ1 lys2Δ0*

EXPERIMENTAL PROCEDURE

▶ Day –3

TA: Patch strains 4-4, 4-5, 4-6, and 4-7 on 16 YPAD plates. Incubate for 3 days at 23°C. Inoculate 5 mL of YPD with strain 4-8 and incubate overnight at 30°C.

▶ Day –2

TA: Transfer strain 4-8 to SPO at 23°C as described in Techniques and Protocols 9 (Sporulation and Tetrad Dissection).

▶ Day 1

Note: *This experiment is to be done INDEPENDENTLY! Each student is responsible for dissection and analysis of their own tetrads.*

Read the Experiment IV Introduction to become familiar with meiosis and tetrad analysis.

Set up a mating between strains 4-6 and 4-7. To do this, use a sterile stick or tooth pick to transfer a small glob of strain 4-6 (~ the size of a 1 mm diameter colony) to a YPD plate. Add 5 µL of sterile H_2O. Transfer a small glob of strain 4-7, using a new sterile stick, and mix the two strains gently together on the surface of the plate. Incubate the YPD plate at 23°C overnight.

▶ Day 2

Streak your mating for single colonies on SD-Ade-Trp plates in order to isolate diploids. Diploids can also be isolated from a mating mixture by micromanipulation, known as "pulling zygotes," which we will do on Day 12. Incubate for 2 days at 30°C.

Media note: *In this experiment, we use SD dropout plates in which the dropout mix is CSM, which differs from the SC dropout mix used in SC plates. See Appendix A for details.*

▶ Days 3 through 9

Meiosis and sporulation are usually complete within 4 days following introduction of diploid cells to sporulation media. A sporulated culture contains a mixture of unsporulated diploid cells, asci with four haploid spores, and asci with fewer than four spores. Spores will not germinate or divide on sporulation media, and most sporulated cultures can be stored in the refrigerator for several months with only a gradual loss of viability.

The instructors have started sporulating 4–8 diploids 3 days ago (you will also sporulate your own versions of strain 4-8 on Day 5, but this will help you get a start on dissecting). During Days 3 through 9, *each student* will need to produce 20 dissected tetrads with four viable spores from strain 4-8. Although you may think this is difficult, particularly with the first few tetrads, you will quickly improve. You can use cells provided by the TA or your own sporulated cells.

Monitor sporulation by removing 5 µL of culture and placing it on a microscope slide. If sporulation was done on plates, scrape a few cells off the plate with a toothpick and resuspend in 5 µL of H_2O on a microscope slide. Top with a coverslip to examine the sporulated culture under a microscope. Identify the unsporulated cells, the four-spored asci and the asci with fewer than four spores. We have to wait until at least 30%–40% of the culture forms four-spored asci, which will make your first tetrad dissection experience simpler. As you become better dissectors, you will be able to use cultures that have as few as 2%–5% four-spored asci.

When your cells are sporulated, treat them with zymolase and put them onto a YPAD plate as described in Techniques and Protocols 9 (Sporulation and Tetrad Dissection). We will be using the SporePlay microscope from Singer Instruments to dissect tetrads. This dissection scope is equipped with a fiber-optic needle. Construction of other types of tetrad dissection needles is described in Techniques and Protocols 10 (Making a Tetrad Dissection Needle). Instructors and TAs are available to help you get started dissecting tetrads. Use the note sheets provided in Appendix B to keep track of your progress when dissecting as this will help with your analysis of meiotic progeny.

▶ *Day 4*

Pick four potential diploids and inoculate each into 2 mL of YPD media in a sterile culture tube. Grow them overnight at 30°C.

▶ *Day 5*

As described in Techniques and Protocols 9 (Sporulation and Tetrad Dissection), centrifuge 1 mL of each diploid in a sterile Eppendorf tube. Wash two times in 1 mL of sterile ddH$_2$O. Resuspend in 500 µL of liquid sporulation media. Transfer to sterile culture tubes containing 2 mL of liquid sporulation media. Make sure that the sporulation media is visibly cloudy. Grow at 23°C for 2–4 days until the cells have sporulated (you can check by putting 5 µL on a slide and looking through a light microscope).

▶ *Day 9*

This is your last day to dissect tetrads! Remember, each student needs 20 full four-spore tetrads.

▶ *Day 12*

Tetrads: We will be replica plating our tetrads. If your tetrads are on one to two plates, you can replica directly from the dissection plates. If your tetrads are on multiple plates, see the TA to use the frogger instead of replica plating to analyze your tetrads.

First, prepare your lawns for the mating-type test. Put 500 µL of YPD media into two sterile Eppendorf tubes. Use a P200 tip (or a toothpick or an applicator stick) to put a matchtip size amount of strain 4-4 in one tube, twirling the tip to release the cells. Spread 200 µL of cell suspension on a YPAD plate using glass beads. Repeat this process with another YPAD plate using strain 4-5. Make sure that each plate is labeled with the name of the mating-type test strain. Wait until these plates dry before proceeding.

Replica-plate your tetrads onto the following plates in the order indicated: YPAD, YPAD, SD-Ura, SD-Trp, SD-Ade, then the YPAD plate containing the a-lawn. *Remove* the velvet, *replace* with a new velvet, and replica-plate to the α-lawn. Incubate one YPAD plate overnight at 37°C. Incubate the remaining plates overnight at 23°C.

Note: The keys to successful replica plating are the following: (1) mark the orientation and type of each plate, (2) do not mash the plates down onto the velvet—an even gentle touch will be sufficient and is more effective, (3) follow instructions regarding the order of plates and remember to switch the velvet in between the mating lawns, (4) if you are working with flocculent strains or transfer a large clump of cells due to mashing, the simplest way to remove big clumps from a plate is to simply replica-plate to a clean velvet—enough cells will remain on the plate so that you can score your phenotype, and (5) do not overgrow the tetrad plates—aim for 1-mm colony size.

If you are using the frogger method, put 100 µL of YPD into all of the wells of a 96-well plate using a multichannel pipette. Use toothpicks to transfer your spore colonies into individual wells, being sure you keep track of the tetrad relationships (each tetrad should be a column [1, 2, 3, etc., through 8], with the four spores being A, B, C, D). Use the frogger to transfer cells to YPAD, YPAD, SD-Ura, SD-Trp, SD-Ade, then the YPAD plate containing the a-lawn. Rinse off the frogger and rest-erilize. After cooling, pin to the α-lawn. Incubate one YPAD plate overnight at 37°C. Incubate the remaining plates overnight at 23°C.

Zygotes: We will learn how to isolate diploids using a micromanipulator. Set up a mating between strains BY4741 and BY4742. To do this, transfer a clump of cells from each strain (~the size of a 1-mm diameter colony) from the plate provided by the TA (these are fresh patches of the strains, on YPAD, grown overnight at 30°C) and mix on a YPAD plate in 5 µL of sterile H_2O, using a stick. Incubate the YPAD plate at 30°C.

After 3 h, transfer a small amount of cells (barely visible on a P20 tip) to an Eppendorf tube containing 100 µL of sterile dH_2O. Transfer 10 µL to a YPAD plate, and tilt the plate so that the drop spreads along a line down the center of the plate (as if you were going to dissect tetrads). Use a SporePlay tetrad dissection microscope to move 5–10 zygotes in a line at position "C." Zygotes have a distinctive appearance, particularly when they re-bud for the first time (Fig. 5). Patch a tiny amount of BY4741 and BY4742 *below* the line of the mating mixture (these are your controls for the replica plating, and you do not want them on top of your zygotes). If you do not have any zygotes at 3 h, try again after 4 h.

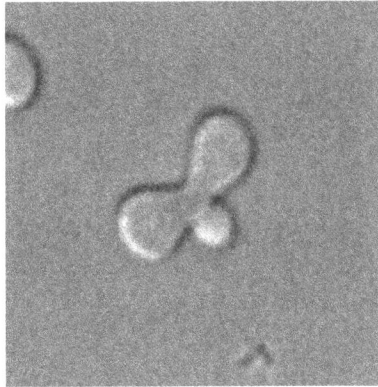

FIGURE 5. DIC image of a zygote that has re-budded.

▶ *Day 13*

Tetrads: Replica-plate your mating test plates to SDmin. Grow overnight at 30°C.

▶ *Day 14*

Tetrads: Score your tetrads using sheets provided in Appendix B. Determine the number of PD, NPD, and T tetrads for each pairwise combination of markers segregating in your cross. Look at each pairwise combination of genes: 37-ade, trp-37, ura, MAT-trp, MAT-37, trp-ade

Determine from your tetrad data if the genes are linked, completely unlinked, or centromere-linked. The instructors will collect each student's data for combined analysis.

Next, determine the linkage distances for all of the linked genes. If two genes are linked and on the same chromosome, calculate the distance between them by a formula that takes double crossovers into account. NPDs are the only accurate predictor of double crossovers because T tetrads can arise by both single and double crossover events. So,

$$\%\,\text{recombination} = 50 \times \frac{T + 6(\text{NPD})}{\text{Total tetrads}}$$

1% recombination is expressed as the genetic map unit centimorgan (cM).

For centromere-linked genes, this formula cannot be used because NPDs arise from spindle attachments in the first meiotic division, not from double crossovers. Instead, T asci arise due to crossing over between the gene and the centromere. So,

$$\% \, \text{recombination} = 50 \times \frac{\text{(Tetratype tetrads)}}{\text{(Total tetrads)}}$$

By just looking at the segregation of markers, do you have strains that contain both *SPC42-mCherry::URA3* and *NUP49-mTurquoise::URA3*? What kind of tetrad is this? Use the microscope to verify that you have Nup49-mTurquoise Spc42-mCherry strains. Be sure to analyze both an NPD and a T tetrad.

Zygotes: We will perform a mating-type test (*MAT***a**/*MAT*α diploids will not mate with either tester), and replica-plate the putative diploids to SD-met-lys to confirm that the cells that you pulled were indeed zygotes. First, prepare your lawns for the mating-type test like you did on Day 12. Replica-plate your zygotes to an SD-Met-Lys plate (diploids will be met + lys+, whereas the haploids are met + lys– and met–lys+), and to the *MAT***a** and *MAT*α mating-type test plates in the following order:

SD-Met-Lys, then the YPAD plate containing the **a**-lawn (4-4). *Remove* the velvet, replace with a *new* velvet, replica-plate to the α-lawn (4-5).
Incubate overnight at 30°C.

▶ Day 15

Tetrads: Be prepared to present your data. You will be asked to provide the number of PD:NPD:T for each combination of loci and the distance between your linked genes. Be sure to look up the position of genes in SGD so you can also report how well your calculations agree with the positions reported in the database. You and your partner can work together on your data analysis.

Zygotes: Replica-plate the mating-type test plates to SDmin and incubate overnight at 30°C

▶ Day 16

Zygotes: Score the SD-Met-Lys and the mating-type test plates.

MATERIALS

Note: *Amounts provided are the requirements for each student.*

Day −3   Sporulated culture of strain 4-8
                 1 YPAD plate with strains 4-4, 4-5, 4-6, 4-7 streaked for singles
                 YPD medium

Day −2   SPO medium

Day 1   1 YPAD plate
                  Sterile H$_2$O
                  Toothpicks

Day 2   1 SD-Trp-Ade plate

Days 3–9   16 Flat YPAD plates for tetrad dissection
                  Zymolyase 100 T (US Biologicals Z1005)
                  1.2 M sorbitol/0.1 M potassium phosphate solution
                  β-Mercaptoethanol
                  Tetrad dissection microscope

Day 4   5 mL of YPD medium
                  4 Sterile culture tubes

Day 5   Sterile distilled H$_2$O
                  10 mL of sporulation media
                  4 Sterile culture tubes

Day 12   1 YPAD plate with BY4741 and BY4742
                  10 mL of YPD
                  10 YPAD plates
                  2 SD-Ura plates
                  2 SD-Ade plates
                  2 SD-Trp plates
                  2 Velvets

Day 13   4 SDmin plates
                  2 Velvets

Day 14   Microscope slides
                  Coverslips
                  1 mL of YPD liquid
                  2 YPAD plates
                  1 SD-Met-Lys plate

Day 15   2 SDmin plates

Isolation and Characterization of Auxotrophic and Temperature-Sensitive Mutants

To obtain large numbers of mutants, yeast cultures are usually treated with mutagens such as ultraviolet (UV) radiation or ethylmethane sulfonate (EMS). These mutagens are remarkably efficient and can induce mutations at a rate of 5×10^{-4} to 1×10^{-2} per gene without substantial killing. Mutants can also be found spontaneously, but the frequency of recovery is too low to be suitable here. In this experiment, we will isolate auxotrophic and temperature-sensitive mutants from EMS-treated yeast.

AUXOTROPHIC MUTANTS

The generation of auxotrophic mutants has been an essential tool for elucidating biosynthetic pathways. Detailed analysis of the mutants have also identified structure–function relationships within the biosynthetic enzymes (e.g., Lingens and Oltmanns 1964; Lindegren et al. 1965). From a practical perspective, the generation of auxotrophic markers was part of the process that made budding yeast a robust laboratory organism. Most mutants were identified by their failure to grow in the absence of an essential nutrient, such as histidine, as we will do here. Other types of mutants, however, were obtained through positive selection. Exposure to α-aminoadipate (αAA), for example, permits selective growth of cells with *lys2* or *lys5* mutations (Chattoo et al. 1979; Zaret and Sherman 1985). Similarly, 5-fluoro-orotic acid (5-FOA) positively selects for cells with *ura3* or *ura5* mutations (Boeke et al. 1986). 5-FOA is now used in a variety of "counter-selection" assays to obtain cells that have lost plasmids or other dispensable DNA regions that contain the *URA3* gene. In this exercise, we will explore the frequencies and phenotypes of 5-FOA-resistant mutants.

MUTANT ENRICHMENT

Various methods have been developed to enrich for mutants after treatment with a mutagen. The underlying principle is to selectively kill actively growing cells under conditions that temporarily block the growth of the desired mutants (Snow 1966; Thouvenot and Bourgeois 1971; Henry et al. 1975; Walton et al. 1979). Nystatin is commonly used for this purpose because it binds ergosterol in the membranes of growing yeast, forming pores that cause leakage of essential metabolites.

TEMPERATURE-SENSITIVE MUTANTS

Many yeast genes specify proteins that participate in indispensable functions (e.g., RNA polymerases, tRNA synthetases, and cell cycle kinases). Mutations that completely destroy the activity of these proteins are lethal. Typically, hypomorphic alleles that retain only partial activity are used to study these genes. A mutation that affects one of these indispensable proteins such that it can function at low temperature but not at high temperature allows temperature-sensitive growth (Hartwell 1967; Pringle and Hartwell 1981). Yeast cells normally grow at 30°C, and thus most temperature-sensitive mutants have been experimentally defined as mutants that live at 23°C (known as the permissive temperature) and die at 36°C–37°C (known as the nonpermissive or restrictive temperature). Temperature-sensitive mutants often grow at intermediate temperatures (e.g., 30°C), albeit sometimes with growth defects. Such intermediate temperatures are therefore referred to as semipermissive temperatures.

COMPLEMENTATION TESTING

A common question that arises during mutant hunts is how many different genes are represented in this collection of mutants? For instance, in a collection of five mutants that cannot grow without leucine, are all of the mutations in the same gene or are the mutations in different genes of the leucine biosynthetic pathway? The most effective way to determine how many of your new mutations occurred in the same gene is to use the complementation test, originally known as the *cis–trans* test (Pontecorvo 1958). Typically, recessive mutations that are in different genes complement one another. Consider two mutants that fail to grow on SC-leu: *leu#1* and *leu#2*. If these are in different genes, then the diploid formed containing both mutants will be *leu#1/LEU LEU/leu#2*, and it will be able to grow on SC-leu because the auxotrophy is "complemented" by the wild-type gene from the other mutant. In this case, *leu#1* and *leu#2* are considered to belong to two different complementation groups. If the mutations are in the same gene, then the diploid formed from both mutants will be *leu#1/leu#2* and it will not be able to grow on SC-leu because there is no functional copy of the gene. This "failure to complement" is the hallmark of mutations in the same gene. In this case, *leu#1* and *leu#2* are said to belong to the same complementation group. Most yeast screens are done in haploids, and thus, the only requirement for complementation testing is the availability of mutants in opposite mating types. This can be achieved by performing half of the mutagenesis screen in *MAT***a** and the other half in *MAT*α, as we will do. Mating type can also be switched using the HO endonuclease (Experiment III). Assigning mutants into complementation groups is useful in terms of cloning and subsequent analysis of mutant phenotypes. Determining the number of complementation groups is also useful to know when to stop screening: when all of the genes involved in a particular process have been identified with multiple mutations.

Complications emerge in complementation testing when there is more than one relevant mutation within a strain or when a single mutation is dominant. Both of these situations can be easily identified by back-crossing to a nonmutagenized parental strain. Another complication emerges when different mutations within the same gene complement one another (intragenic complementation). Genes that encode multiple independent functions, like *HIS4* of the histidine biosynthetic pathway, are susceptible to intragenic complementation (e.g., Fink 1966). A final yet rare complication occurs when recessive alleles of two unlinked genes fail to complement one another. Such cases of nonallelic noncomplementation usually signify that the products of the genes interact with one another physically (e.g., Stearns and Botstein 1988).

EXPERIMENTAL OVERVIEW

The mutagenesis described here involves treatment of wild-type yeast strains with ethylmethane sulfonate (EMS). Half of the class will mutagenize a *MAT***a** strain and half will mutagenize a *MAT*α strain. After mutagenesis, the strains will be diluted and plated onto complete medium plates at a concentration of about 200 cells/plate. After these cells have grown into colonies, they will be transferred to various media by replica plating. Temperature-sensitive mutants will be detected by comparing pairs of plates that were incubated at 23°C and 37°C. Histidine auxotrophs will be detected by comparing plates containing synthetic complete media with those that lack the amino acid (Fink 1964, 1966). Uracil auxotrophs will be detected by growth on media containing the counterselective agent 5-FOA (Boeke et al. 1986).

STRAINS

5-1	MRG5388 *MAT*α *mal gal2*
5-2	yJHK102 *MAT***a** *can1-100*
5-3	BY4741 *MAT***a** *his1*Δ with integrated pXD-URA3,SpHIS5MX
5-4	BY4741 *MAT***a** *his2*Δ with integrated pXD-URA3,SpHIS5MX

5-5	BY4741 *MAT**a** his3Δ*
5-6	BY4741 *MAT**a** his4Δ* with integrated pXD-URA3,SpHIS5MX
5-7	BY4741 *MAT**a** his5Δ* with integrated pXD-URA3,SpHIS5MX
5-8	BY4741 *MAT**a** his6Δ* with integrated pXD-URA3,SpHIS5MX
5-9	BY4741 *MAT**a** his7Δ* with integrated pXD-URA3,SpHIS5MX
5-10	BY4742 *MATα his1Δ* with integrated pXD-URA3,SpHIS5MX
5-11	BY4742 *MATα his2Δ* with integrated pXD-URA3,SpHIS5MX
5-12	BY4742 *MATα his3Δ*
5-13	BY4742 *MATα his4Δ* with integrated pXD-URA3,SpHIS5MX
5-14	BY4742 *MATα his5Δ* with integrated pXD-URA3,SpHIS5MX
5-15	BY4742 *MATα his6Δ* with integrated pXD-URA3,SpHIS5MX
5-16	BY4742 *MATα his7Δ* with integrated pXD-URA3,SpHIS5MX
5-17	BY4741 *MAT**a** ura3Δ*
5-18	BY4741 *MAT**a** ura5Δ* with pXR-URA3,SpHIS5MX
5-19	BY4742 *MATα ura3Δ*
5-20	BY4742 *MATα ura5Δ* with pXR-URA3,SpHIS5MX

SAFETY NOTES

EMS is a volatile organic solvent that is a mutagen and carcinogen. It is harmful if inhaled, ingested, or absorbed through the skin. Your TA will wear gloves and protective clothing, and work in a chemical fume hood when performing the mutagenesis. Disposable tubes and plastic pipettes will be used for all manipulations. A waste container containing 5% sodium thiosulfate will be located in the hood for both solid and liquid EMS waste. Store EMS in the refrigerator until immediately before use to minimize its volatility.

EXPERIMENTAL PROCEDURES

▶ *Day 1*

Half of the class will inoculate 5 mL of YPD with *MATα* strain 5-1. The other half will inoculate with *MAT**a*** strain 5-2. Grow overnight at 30°C. Note which strain you used.

▶ *Day 2*

Making mutants

Using a saturated overnight culture, prepare cells for mutagenesis as described in Techniques and Protocols 11 (EMS Mutagenesis). Determine the cell density of your culture with a hemocytometer (Techniques and Protocols 12). Split the cell sample and give half to your TA who will mutagenize the cells with EMS. While your cells are being treated, label your plates for the subsequent steps of the day.

Isolating mutants

Dilute the EMS-treated culture to ~2000 cells/mL based on prior hemocytometer counts. Assume that no cells were lost in the mutagenesis procedure. This should be on the order of a 1:100,000 dilution, but it may vary depending on the initial concentration of cells used. Spread 0.1, 0.2, and 0.4 mL on separate YPD plates, using 10 plates for each of the three different volumes plated (30 plates total). Incubate all the plates at room temperature (23°C) until Day 6.

Isolating uracil auxotrophs

We will search for mutations in *URA3* and *URA5* using the counterselectable agent 5-FOA in two different ways. In the first case, spread 0.1 mL of mutagenized and nonmutagenized cells (undiluted) directly onto 5-FOA plates. Incubate at 30°C until Day 6.

In the second case, spread 0.1 mL of mutagenized and nonmutagenized cells (undiluted) into YPD plates. Grow overnight at 30°C. These cells will be replica-plated to 5-FOA on the following day.

▶ Day 3

Uracil auxotrophs

Replica-plate the YPD plate of undiluted mutants to 5-FOA plates. Incubate at 30°C until Day 7.

▶ Day 6

Isolating mutants

Count the colonies on your YPD plates, and estimate the survival after EMS. Record this number so that you can present an estimate of your mutagenesis frequency in your report at the end of the course.

Average number of colonies on plates =
Expected number of colonies on plates if no EMS =
Mutagenesis frequency based on lethality (average/expected) =

Choose 10 YPD plates containing ~200 colonies/plate for the isolation of mutants by replica plating. To isolate histidine auxotrophic mutants, replica-plate to SC and SC-his plates and incubate at 30°C. To isolate temperature-sensitive mutants, replica-plate to two YPD plates and incubate one at 23°C and the other at 37°C. Make sure that each plate is numbered and has an orientation symbol on the back.

Uracil auxotrophic mutants

Place 5-FOA plates from Day 2 at 4°C for safekeeping.

▶ Day 7

Histidine auxotrophic mutants

Identify up to 20 individual colonies that grow on SC but not on SC-his. Give each of these mutants a systematic name that includes your group number (e.g., the 10 mutants from group 5 span from *hisA5*, *hisB5*, *hisC5*, … to *hisT5*). Restreak each of your mutants onto SC and SC-his plates to confirm histidine auxotrophy.

Uracil auxotrophic mutants

Count the number of colonies arising on the 5-FOA plates spread directly on 5-FOA or spread after overnight growth on YPD. Record the data in the table below. Using 2×10^8 cells/ml as the density of your starting culture, estimate the frequency of uracil auxotrophy before mutagenesis and after mutagenesis for the two different plating procedures (direct or delayed).

	Direct plating on 5-FOA	Delayed plating on 5-FOA
# of colonies after EMS treatment		
# of colonies without EMS treatment		
Fold increase in colony-forming units due to EMS		
Frequency of uracil auxotrophy		

Did the frequency of FOA resistance increase after mutagenesis? Why might growth after mutagenesis affect the recovery of FOA-resistant mutants? Purify six colonies from a 5-FOA plate by streaking on YPD for subsequent analysis of their Ura phenotype. Incubate the plates at 30°C.

▶ Day 8

Histidine auxotrophic mutants

To prepare for complementation testing, make master plates of your histidine auxotrophs in the following manner. Use a toothpick to spread each mutant into a thin solid line across the surface of a YPD plate (10 mutants/plate, see Appendix D for a template). Use your nonmutagenized parent to create an

11th line. The top and bottom lines should be spaced about 2 cm from the edges of the plate. Repeat to make a second master plate. Incubate both overnight at 30°C.

▶ Day 9

Histidine auxotrophic mutants

Two types of complementation tests will be performed to classify the histidine auxotrophs. In the first type, you will cross each of your mutants to strains with null mutations in genes of the histidine biosynthetic pathway, *HIS1* through *HIS7*. This form of "candidate testing" can greatly accelerate a mutant hunt. Your TA will provide a master plate containing the seven *his1-7* mutations in either *MAT**a*** or *MATα* strains, which you will replica-plate to a fresh velvet. Next, replica-plate one of your master plates to the velvet such that the streaks of the two plates are perpendicular. Print from this velvet onto a fresh YPD plate to allow mating. Repeat if you collected more than 10 mutants. Incubate at 30°C overnight.

Important: *Do not contaminate the master plates because other groups need to use them. Therefore, print the master first and then your own plate to the velvet. Furthermore, use a new velvet for each set of mutants.*

In the second complementation test, we will sort the class-wide collection of mutants into complementation groups by crossing all of the *MAT**a*** mutants to all of the *MATα* mutants. To this end, each group will give the second copy of their master mutant plates to the TA who will create the crosses on YPD. Copies of these cross-streaked plates will be given back to you before the start of the next day.

Uracil auxotrophic mutants

Streak the purified 5-FOA^R colonies onto SC and SC-ura plates to determine whether 5-FOA-resistant colonies are truly auxotrophic for uracil.

Temperature-sensitive mutants

Compare the 37°C plate with the 23°C plate. Pick ~12 colonies that failed to grow at the high temperature and restreak onto two new YPD plates and two YPD plates containing 30% sucrose. Include your parental strain as a control. Incubate one YPD plate and one YPD + sucrose plate at 37°C, and the other set of plates at 23°C for 2 d. The high osmolarity of YPD + sucrose plates induces a cellular response that includes the production of glycerol, which stabilizes some mutant proteins in their active conformations (osmotic remediation). The sucrose test may also identify mutants that cause lysis in high osmotic media.

▶ *Day 10*

Histidine auxotrophic mutants

Replica-plate the cross-streaked mutants of both your complementation tests to SC and SC-his plates. Incubate at 30°C overnight.

Uracil auxotrophic mutants

Record the growth of the *ura3* mutants from Day 9. Mate each of your uracil auxotrophs to known *ura3* and *ura5* null standards of opposite mating type (strains 5-17 and 5-18 or strains 5-19 and 5-20) by patching or cross-streaking on a YPD plate. Incubate at 30°C overnight.

Temperature-sensitive mutants

Record the growth of your mutants at 23°C and 37°C. Assign a number to each mutant. Note the fraction of mutants that respond to sucrose in the media.

▶ *Day 11*

Histidine auxotrophic mutants

Evaluate your complementation tests.

- In the class-wide complementation test, how many cases of noncomplementation did you observe?

- Did you obtain more complementation groups than the number of known genes in the histidine biosynthtic pathway?

- Did any of your isolates fail to fit easily into a single complementation group?

- In the case of candidate testing, which of your mutants correspond to known genes?

- Did any isolates that are complemented by *HIS4* also complement one another?

Uracil auxotrophic mutants

Replica-plate the complementation tests to SC and SC-ura.

▶ *Day 12*

Uracil auxotrophic mutants

Score your complementation tests.

- How many isolates contain mutations in *URA3*, and how many contain mutations in *URA5*?

- How do you explain mutations that are not complemented by either the *ura3* or *ura5* standards?

MATERIALS

Note: *Amounts provided are the requirements for each pair of students.*

Day 1	1 Culture tube with 5 mL of YPD
	Master plate with strains 5-1 and 5-2
Day 2	10 mL of sterile distilled H_2O
	32 YPD plates
	2 5-FOA plates
Day 3	2 5-FOA plates
	Velvets
Day 6	10 SC plates
	10 SC-his plates
	20 YPD plates
	Velvets
Day 7	2 SC plates
	2 SC-his plates
	1 YPD plate
Day 8	4 YPD plates
Day 9	6 YPD plates
	1 SC-ura plate
	1 SC plate
	2 YPD plates containing 30% sucrose
	Master plate with strains 5-3 to 5-9
	Master plate with strains 5-10 to 5-16
	Velvets
Day 10	2 SC plates
	2 SC-his plates

1 YPD plate
Velvets

Day 11   1 SC-ura plate
1 SC plate

REFERENCES

Auxotrophic Mutants

Lindegren G, Hwang LY, Oshima Y, Lindegren C. 1965. Genetical mutants induced by ethyl meth-anesulfonate in *Saccharomyces*. *Can J Genet Cytol* **7:** 491–499.

Lingens F, Oltmanns O. 1964. Erzeugung und untersuchung Biochemischer and Mangelmu-tanten von *Saccharomyces cerevisiae*. *Z Naturforsch* **19B:** 1058–1065.

Lingens F, Oltmanns O. 1966. Uber die Mutagene Wirkung von 1-nitroso-5-nitro-1-methyl-guanidin (NNMG) und *Saccharomyces cerevisiae*. *Z Naturforsch* **21B:** 660–663.

Temperature-Sensitive Mutants

Hartwell LH. 1967. Macromolecule synthesis in temperature-sensitive mutants of yeast. *J Bacteriol* **93:** 1662–1670.

Pringle JR, Hartwell LH. 1981. The *Saccharomyces cerevisiae* cell cycle. In *The molecular biology of the yeast Saccharomyces: Life cycle and inheritance* (ed. JN Strathern, et al.), pp. 97–142. Cold Spring Harbor Laboratory, Cold Spring Harbor, New York.

Selections for Auxotrophs

Boeke JD, LaCroute F, Fink GR. 1986. A positive selection for mutants lacking orotidine-5'-phosphate decarboxylase activity in yeast; 5'-fluoro-orotic acid resistance. *Mol Gen Genet* **197:** 345–346.

Chattoo BB, Sherman F, Azubalis DA, Fjellstedt TA, Mehnert D, Ogur M. 1979. Selection of *lys2* mutants of the yeast *Saccharomyces cerevisiae* by the utilization of α-aminoadipate. *Genetics* **93:** 51–65.

Zaret KS, Sherman F. 1985. α-Aminoadipate as a primary nitrogen source for *Saccharomyces cere-visiae* mutants of yeast. *J Bacteriol* **162:** 579–583.

Enrichment Methods

Henry SA, Donahue TF, Culbertson MR. 1975. Selection of spontaneous mutants by inositol star-vation in *Saccharomyces cerevisiae*. *Mol Gen Genet* **143:** 5–11.

Snow R. 1966. An enrichment method for auxotrophic yeast mutants using the antibiotic "nys-tatin." *Nature* **211:** 206–207.

Thouvenot DR, Bourgeois CM. 1971. Optimization de la selection de mutants de *Saccharomyces cer-evisiae* par la nystatine. *Ann Inst Pasteur* **120:** 617–625.

Walton BF, Carter BLA, Pringle JR. 1979. An enrichment method for temperature-sensitive and auxotrophic mutants of yeast. *Mol Gen Genet* **171:** 111–114.

Complementation Testing

Fink GR. 1966. A cluster of genes controlling three enzymes in histidine biosynthesis in *Saccharomyces cerevisiae*. *Genetics* **53**: 445–459.

Hawley RS, Walker MY. 2003. *Advanced genetic analysis: Finding meaning in a genome*. Blackwell Publishing, Massachusetts Malden.

Pontecorvo G. 1958. *Trends in genetic analysis*. Columbia University Press, New York, New York.

Stearns T, Botstein D. 1988. Unlinked noncomplementation: Isolation of new conditional-lethal mutations in each of the tubulin genes of *Saccharomyces cerevisiae*. *Genetics* **119**: 249–260.

Working with Essential Genes

Approximately 1000, or ~19%, of *Saccharomyces cerevisiae* genes are considered to be essential, meaning that a haploid deletion mutant is not viable on YPD medium at 30°C. Essential genes are particularly challenging to study in phenotypic assays because mutants that are the simplest to construct, either complete gene deletions or complete loss of function alleles, render the cells inviable. In this experiment, we will discuss methods used to study the function of essential genes.

Classic examples of essential genes are those isolated by Lee Hartwell and colleagues in the famous cell division cycle (*cdc*) mutant screen that identified genes involved in basic cellular functions such as DNA replication and mitosis (Hartwell et al. 1970) and led to the Nobel Prize in 2001. The *cdc* mutants were isolated by mutagenizing a wild-type strain and then characterizing a set of temperature-sensitive (ts) mutants that had cell cycle phenotypes. These mutants were viable at 23°C (the permissive temperature), but inviable when the temperature was shifted to 37°C (the restrictive temperature). The mutant strains can be maintained at low temperature, but they can be studied in phenotypic assays by shifting the temperature to 37°C when needed. Sometimes, the ts mutants can be grown at a "semipermissive" temperature such as 30°C, where the strain can grow well enough but still displays a mutant phenotype for the process that is being studied (e.g., transcription, replication, secretion, etc.). Sometimes, mutants can be cold sensitive (cs), but the principle of shifting from a permissive temperature to a restrictive temperature is the same. It is important to keep in mind that there are only a few examples of conditional alleles where the mechanistic basis of the restrictive condition is known with any certainty, and many conditional alleles remain hypomorphic (i.e., they have less than wild-type function) at the permissive condition.

TEMPERATURE-SENSITIVE MUTANTS

Historically, ts mutants were isolated in phenotypic screens similar to your screen in Experiment V. The types of alleles recovered depended on the mutagenic agent used. For example, ethylmethane sulfonate induces point mutations, particularly base transitions from G:C to A:T, whereas UV-induced alleles often contain small deletions or C to T transitions due to pyrimidine dimer formation. Depending on the position of the mutation, these changes generally result in substitution of a single amino acid. How a single-amino-acid change results in a protein that functions at 23°C but not at 37°C varies from mutant to mutant. In some cases, the protein is destabilized as a result of the mutation causing it to be denatured and degraded at elevated temperatures. Other ts mutants have no effect on protein stability but instead impact protein expression, protein–protein interactions, or posttranslational modifications. It is possible to isolate multiple alleles of a single gene, known as an allelic series. If the gene has multiple functions or distinct binding partners, it is often possible to identify mutants that affect one process but not another, thus "separating" the functions of the gene.

Several strategies exist for generating ts alleles systematically in high throughput (Ben-Aroya et al. 2010). These typically involve a polymerase chain reaction (PCR)-based mutagenesis followed by transformation and in vivo recombination, but invariably require screening many potential mutants.

POINT MUTANTS

In some instances, a specific mutation is desired. Examples include changing sites of posttranslational modifications and identifying functional regions of a protein. Non phosphorylatable or phosphomimetic alleles are made by changing serine or threonine residues to alanine or aspartic/glutamic acid, respectively. Important regions of a protein can be identified or characterized by making small deletions or by performing a scanning mutagenesis in which every residue is individually changed to alanine. This latter approach is often impractical for large proteins. Although it is possible to make deletions, these can interfere with protein folding, rendering the protein nonfunctional. A more recent strategy for mutagenesis takes advantage of the wealth of sequence data within the ascomycetes. A good starting point for mutagenesis of an uncharacterized protein is to alter residues that Nature has deemed important and therefore have remained highly conserved throughout evolution. There are several strategies for the creation of point mutations and deletions in a gene as well as for introducing the allele into yeast. The two most common methods are site-directed mutagenesis of a gene cloned into a plasmid and introduction into yeast using the plasmid shuffle, and PCR-mediated mutagenesis and introduction into yeast by one-step gene replacement.

GENERATION OF CONDITIONAL ALLELES USING PROMOTER SHUT-OFF, N-DEGRON, AND AUXIN-INDUCIBLE DEGRON

Today, alleles of a gene are often created in the lab and introduced into yeast using PCR-derived fragments or plasmids. Several reverse genetics approaches for designing conditional alleles of an essential gene can be used and are often chosen based on the question being studied. Conditional alleles can be constructed by fusing the gene of interest to a regulated promoter, such that transcription of the gene can be turned off at will. Three conditional expression systems are commonly used in yeast: the Tet-off system and the *MET3* and *GAL1* promoters. For example, to generate a conditional mutant strain, the endogenous promoter is replaced by the *GAL1* promoter. When such a strain is grown on galactose-containing medium, the gene is highly expressed. The *GAL1* promoter is tightly repressed in the presence of glucose, and thus, expression of the gene can be repressed by shifting the cells to glucose-containing medium. The availability of yeast open reading frame (ORF) libraries in *GAL1* promoter vectors (Gelperin et al. 2005; Douglas et al. 2012) facilitates the generation of conditional strains with *GAL1*pr-ORF covering deletion of the given ORF. The *MET3* promoter offers an alternative to regulating expression by carbon source, and it is less active than the *GAL1* promoter in its "on" state, resulting in a lower expression level that in some cases can be more readily shut off (Black et al. 1995). The Tet-off system (Gari et al. 1997; Mnaimneh et al. 2004) uses an artificial promoter that contains binding sites for a Tet repressor: Transcriptional activator fusion (tTA*). In normal growth medium, tTA* is bound to the artificial promoter and transcription of the gene is induced. Addition of tetracycline (or doxycycline, a tetracycline analog) dissociates tTA* from the promoter, turning transcription off. In all cases, mutant phenotypes can then be detected following the shift to the promoter "off" condition. This kind of experiment requires turnover of the existing gene products, and thus works best for proteins with short half-lives.

A second approach targets the stability of the protein, rather than transcription, and results in the regulated proteolysis of the protein of interest. Degradation can be induced by high temperature (N-degron) or by a small molecule (auxin-inducible degron). The N-degron method (Dohmen et al. 1994; Kanemaki et al. 2003) fuses the amino terminus of the protein of interest to ubiquitin followed by a modified dihydrofolate reductase (DHFR) that unfolds at elevated temperature. Following translation, the ubiquitin is rapidly cleaved, which uncovers an amino-terminal arginine residue of DHFR. The ubiquitin ligase, Ubr1, recognizes the arginine and, at 37°C, targets the fusion protein for polyubiquitination and subsequent degradation due to partial unfolding of the DHFR domain. The auxin-inducible degron (AID) system (Nishimura et al. 2009; Morawska and Ulrich 2013) provides an alternative to the use of high temperature to induce protein degradation. In this method, an AID-tag is added to the protein of interest in a strain that expresses the Tir1 F-box protein of plants. In the presence of plant hormone 1-naphthalenacetic acid (NAA), Tir1 binds the AID-tag, as well as the yeast Skp1-Cullin-F-box (SCF) E3 ligase, which results in polyubiquitination and proteosomal destruction of the targeted protein. Both degron approaches have been particularly useful for reversibly arresting the cell cycle by conditionally degrading key regulatory proteins, then allowing cells to reenter cell division following removal of the hormone (AID) or shift back to 23°C (N-degron).

In this experiment, we will compare the different systems of conditional alleles, including classic *CDC* ts alleles, *GAL1* promoter strains, Tet-off strains, N-degron strains, and AID strains. We will compare growth, morphology, and cell cycle position of strains carrying conditional alleles of the CDK catalytic subunit gene *CDC28* and of the DNA replication protein *MCM4*.

STRAINS

6-1 BY4741	*MATa his3Δ1 leu2Δ0 ura3Δ0 met15Δ0*	Control
6-2	*MATa cdc28-1::kanMX his3Δ1 leu2Δ0 ura3Δ0 met15Δ0*	ts
6-3	*MATa cdc28-4::kanMX his3Δ1 leu2Δ0 ura3Δ0 met15Δ0*	ts
6-4	*MATa cdc28-13:kanMX his3Δ1 leu2Δ0 ura3Δ0 met15Δ0*	ts
6-5	*MATa cdc28-td::kanMX his3Δ1 leu2Δ0 ura3Δ0 met15Δ0*	N-degron
6-6	*MATa mcm4-PH::kanMX his3Δ1 leu2Δ0 ura3Δ0 met15Δ0*	ts
6-7	*MATa ura3-1::ADH1pr-OsTIR1-9Myc:: URA3MX* *ade2-1 his3-11,15 leu2-3,112 trp1-1 can1-100*	
6-8	*MATa ura3-1::ADH1pr-OsTIR1-9Myc:: URA3MX* *mcm4::MCM4-AID::kanMX* *ade2-1 his3-11,15 leu2-3,112 trp1-1 can1-100*	AID
6-9	*MATa URA3::CMV-tTA his3Δ1 leu2Δ0 met15Δ0*	Control (Tet)
6-10	*MATa URA3::CMV-tTA his3Δ1 leu2Δ0 met15Δ0* *mcm4::kanMX::tetO7-TATA-MCM4*	Tet-off
6-11	*MATa his3Δ1 leu2Δ0 ura3Δ0 met15Δ0* pBY011-D123[URA3MX]	Control (GAL1)
6-12	*MATa cdc28Δ::kanMX his3Δ1 leu2Δ0 ura3Δ0 met15Δ0* pBY011-CDC28[URA3MX]	*GAL1* promoter

6-13   *MAT**a** mcm4Δ::kanMX his3Δ1 leu2Δ0 ura3Δ0*   *GAL1* promoter
              met15Δ0
              pBY011-*MCM4[URA3MX]*

EXPERIMENTAL PROCEDURE

▶ Day 1

Streak out *all* strains for single colonies using the method described in class. Use YPAD plates for strains 6-1 to 6-10. For strain 6-11 to 6-13, grow strains on SGR-Ura. The galactose and raffinose in the media will induce expression from the *GAL1* promoter and the Ura selection will ensure that the plasmid is maintained in each strain. Incubate plates at 23°C for 2 days.

▶ Day 3

Lightly inoculate 5 mL of YPAD (for strains 6-1 to 6-10) or SGR-Ura (for strains 6-11 to 6-13) in culture tubes. Incubate cultures at 23°C overnight with agitation.

▶ Day 4

First thing in the morning, measure the OD_{600} of cultures 6-1 to 6-6. The OD_{600} should be greater than 1.0. Dilute a new 5-mL YPAD culture of each to OD_{600} of 0.1, label these strains as "log" strains (meaning logarithmically growing), and return all cultures to incubate at 23°C for 4–6 h. Incubate 50 mL of YPAD media at 37°C for use in the time course later.

For time course: Measure the OD_{600} of each logarithmically growing strain that you diluted this morning (strains 6-1 to 6-6). Dilute strains in 6 mL (final volume) of YPAD to $OD_{600} = 0.3$ in 15-mL conical tubes.

Place a 1-mL sample in an Eppendorf tube. Centrifuge at maximum speed for 1 min. Remove the media and fix cells in 1 mL of 70% ethanol.

Centrifuge the remaining cells in the 15-mL conical tube at 3000 rpm for 3 min at room temperature. Pour off media. Add 5 mL of 37°C YPAD to each culture. Resuspend the cells by gentle vortexing. Transfer the cells into new culture tubes. Incubate at 37°C for 3 h in a shaking water bath. Place another 1-mL sample in an Eppendorf tube, centrifuge at maximum speed for 1 min, aspirate, and fix in 1 mL of 70% ethanol.

For spot assay: Measure the OD_{600} of each saturated culture (strains 6-1 to 6-13). For *GAL1* strains (6-11 to 6-13), wash the cells *before* plating. Place 1 mL of cells into an Eppendorf tube. Centrifuge at 5000 rpm for 2 min. Aspirate the media and add 1 mL of dH_2O. Resuspend cells by gentle vortexing. Repeat centrifugation, aspirate, and resuspend cells in dH_2O. Use washed cells for the remainder of the experiment.

Serial dilutions will be made in a 96-well plate. A 96-well pin tool will be used to transfer cells onto each OMNI (rectangle) plate. To use the pin tool, it is important that plates are *dry* or else the spots of each culture will not absorb into the media properly.

Calculate the volume of cells needed for an OD_{600} of 0.5 in a final volume of 200 µL, using the OD_{600} that you measured for each saturated culture. Add the appropriate volume of each culture into rows A and E and add YPAD (for strains 6-1 to 6-10) or dH_2O (for strains 6-11 to 6-13) to bring the final volume to 200 µL. Use a multichannel pipette to add 180 µL of dH_2O to the remaining wells of the 96-well plate. Next, use the multi-channel pipette to mix the wells containing cells and transfer 20 µL into the wells below it to make a 1 in 10 dilution. Mix and repeat two more times to make 1 in 100 and 1 in 1000 dilutions.

Place pin tool into thin layer (~5 mm) of 70% ethanol and let it soak for 10 sec. Shake off excess ethanol and quickly pass the pins through the flame.

Note: The pin tool will catch fire to sterilize. It is VERY IMPORTANT that you keep the pins horizontal, but with a slight angle toward the ceiling away from you (you probably do not want to light yourself on fire, nor do you want flaming ethanol to run down the pin tool onto your arm). Allow the pin tool to cool for 1 min.

Place the pin tool into the 96-well plate. Stir gently for 5 sec and then let the pin tool sit on the bottom of the plate for 10 sec. Place the pin tool onto an OMNI plate for 10 sec. Repeat for all seven plates. You *do not* need to resterilize the pin tool between plates. Just ensure that you keep the same orientation for the pin tool.

Incubate your plates in the appropriate temperature:

Plate type	Incubation temperature (°C)
YPAD	23
YPAD	30
YPAD	37
YPAD + 10 µg/mL Doxycycline	30
YPAD + 500 µM NAA	30
SGR-Ura	30
SD-Ura	30

▶ Day 6

Take a look at your YPAD plate at 30°C. Because this plate is being grown at permissive temperature on rich media lacking the drug, the strains may grow faster on this plate in comparison to the others. If this seems to be the case, incubate this plate at 4°C to slow down growth until tomorrow.

▶ *Day 7*

For spot assay: Take out plates, keep them at room temperature for 30 mins and than scan your plates. Record your findings for each plate. Be sure to compare each mutant to its appropriate control strain.

For time course: Split your ethanol-fixed cells into two 500-µL aliquots. One aliquot will be used for analysis of the DNA content by flow cytometry. The other aliquot will be used to examine cell morphology and budding index microscopically.

Prepare one 500-µL sample from each time point for flow cytometry as detailed in Techniques and Protocols 13 and leave the cells in FACS buffer. SYBRGreen, which is light sensitive, should not be added to cells until immediately before running on the C6 cytometer (tomorrow). If you are unable to use the C6 tomorrow, store cells in FACS buffer at 4°C until you are ready. Samples are stable in FACS buffer at 4°C for ~1 wk.

▶ *Day 8*

Complete the flow cytometry protocol by adding SYBRGreen and sonicating your samples. Run your prepared flow cytometry samples on the C6. For instructions on how to run the C6, see Techniques and Protocols 3.

▶ *Day 10 to 12*

Look at your flow cytometry data. Do you see cell cycle differences between the different shut-off strains?

Choose one or two conditional strains that you would like to examine. Along with the appropriate wild-type control(s), centrifuge the 500-µL aliquot of ethanol-fixed samples for the 0 h and 3 h time points at max speed for 1 min (this will give you four to six samples in total). Aspirate the ethanol and add 10–20 µL of mounting medium with DAPI (VECTASHIELD). Resuspend cells by gently pipetting up and down. Spot 3 µL of the medium onto a glass slide, apply a coverslip, and image on a wide-field fluorescence microscope. It will take ~15 min to make your slides, and ~30–60 min to image four slides and collect micrographs.

▶ *Day 15*

Rank your conditional alleles. Which shut-off alleles worked the best in the spot assay? How did the alleles from the time course affect the cell cycle? Did any of the shut-off strains have abnormal morphology? Do the phenotypes seem to be logical given the biological roles for Cdc28 and Mcm4 in the cell?

MATERIALS

Day 1   2 YPAD plates
             1 SGR-Ura plate
             Sticks

Day 3   Culture tubes
             YPAD
             SGR-Ura
             Sticks

Day 4   Cuvettes
             Culture tubes
             YPAD
             50-mL Conical tub
             15-mL Conical tubes
             Eppendorf tubes
             70% Ethanol
             Sterile dH_2O
             96-well plate
             96-well pin tool
             Multichannel pipette
             3 YPAD OMNI plates
             1 YPAD + 10 µg/mL Doxycycline OMNI plate
             1 YPAD + 500 µM NAA OMNI plate
             1 SGR-Ura OMNI plate
             1 SD-Ura OMNI plate

Day 7   Scanner
             dH_2O
             50 mM Tris-Cl (pH 8.0)
             RNase A
             50 mM Tris-Cl (pH 7.5)
             Proteinase K
             FACS buffer (see recipe in Techniques and Protocols 13)

Day 8   50 mM Tris-Cl (pH 7.5)
             SYBRGreen
             Sonicator

Day 10 Mounting media containing DAPI
Microscope slides and coverslips

REFERENCES

Ben-Aroya S, Pan X, Boeke JD, Hieter P. 2010. Making temperature-sensitive mutants. *Methods Enzymol* **470:** 181–204.

Black S, Andrews PD, Sneddon AA, Stark MJ. 1995. A regulated MET3-GLC7 gene fusion provides evidence of a mitotic role for *Saccharomyces cerevisiae* protein phosphatase 1. *Yeast* **11:** 747–759.

Dohmen RJ, Wu P, Varshavsky A. 1994. Heat-inducible degron: A method for constructing temperature-sensitive mutants. *Science* **263:** 1273–1276.

Douglas AC, Smith AM, Sharifpoor S, Yan Z, Durbic T, Heisler LE, Lee AY, Ryan O, Gottert H, Surendra A, et al. 2012. Functional analysis with a barcoder yeast gene overexpression system. *G3 (Bethesda)* **2:** 1279–1289.

Gari E, Piedrafita L, Aldea M, Herrero E. 1997. A set of vectors with a tetracycline-regulatable promoter system for modulated gene expression in *Saccharomyces cerevisiae*. *Yeast* **13:** 837–848.

Gelperin DM, White MA, Wilkinson ML, Kon Y, Kung LA, Wise KJ, Lopez-Hoyo N, Jiang L, Piccirillo S, Yu H, et al. 2005. Biochemical and genetic analysis of the yeast proteome with a movable ORF collection. *Genes Dev* **19:** 2816–2826.

Hartwell LH, Culotti J, Reid B. 1970. Genetic control of the cell-division cycle in yeast. I. Detection of mutants. *Proc Natl Acad Sci* **66:** 352–359.

Kanemaki M, Sanchez-Diaz A, Gambus A, Labib K. 2003. Functional proteomic identification of DNA replication proteins by induced proteolysis *in vivo*. *Nature* **423:** 720–724.

Mnaimneh S, Davierwala AP, Haynes J, Moffat J, Peng WT, Zhang W, Yang X, Pootoolal J, Chua G, Lopez A, et al. 2004. Exploration of essential gene functions via titratable promoter alleles. *Cell* **118:** 31–44.

Morawska M, Ulrich HD. 2013. An expanded tool kit for the auxin-inducible degron system in budding yeast. *Yeast* **30:** 341–351.

Nishimura K, Fukagawa T, Takisawa H, Kakimoto T, Kanemaki M. 2009. An auxin-based degron system for the rapid depletion of proteins in nonplant cells. *Nat Methods* **6:** 917–922.

Synthetic Lethal Mutants and Random Sporulation

One of the advances resulting from the yeast genome project is a complete set of deletion mutants. A consortium of labs collaborated to construct deletion mutations of every gene in the BY4743 background. This strain background is largely isogenic with S288c, but it does have some differences (e.g., it contains an allelic variant of *MIP1*, which causes an increased frequency of petite mutants). The BY4743 strain contains homozygous *ura3Δ0*, *his3Δ1*, and *leu2Δ0* alleles and heterozygous *lys2Δ0* and *met15Δ0* alleles (see http://www-sequence.stanford.edu/group/yeast_deletion_project/deletions3.html).

The *ura3Δ0*, *leu2Δ0*, *lys2Δ0*, and *met15Δ0* alleles are complete deletions, and the *his3Δ1* allele is a 187-bp internal deletion of the *HIS3* gene (Giaever et al. 2002). The deletions of all other open reading frames (ORFs) were constructed by one-step gene replacements using polymerase chain reaction (PCR)-derived fragments and the G418 resistance (G418^r) cassette from the pFA6a family of plasmids. Each deletion was constructed in a diploid so that the results were heterozygous strains in which one copy of the gene was deleted. Correct integrations were confirmed by both PCR and tetrad analysis. Two general classes of genes are defined by the deletions. Nonessential genes are defined by deletion mutants that contain four viable spores upon tetrad dissection, two of which are G418^r. Essential genes are defined by mutants that produce two viable spores, neither of which are G418^r.

The deletion strains can be purchased from vendors such as GE Healthcare Dharmacon (http://dharmacon.gelifesciences.com/non-mammalian-cdna-and-orf/yeast-knockout-collection/), transOMIC Technologies (http://www.transomic.com/Products/Yeast-Products/Yeast-Knock-Out-Collection/Product-Overview.aspx), ATCC (http://www.atcc.org/Products/Cells_and_Microorganisms/Fungi_and_Yeast/Saccharomyces_cerevisiae_Deletion_Mutants.aspx), and EUROSCARF (http://www.euroscarf.de). Storage and handling of the deletion collection are detailed in Techniques and Protocols 14. Complete sets with deletions of nonessential genes are available as haploids and diploids, and the complete set of deletions (essential and nonessential genes) is also available as heterozygous diploids. Precise deletions of all genes add two important tools to the repertoire of yeast genetics. First, the collection can be screened in a systematic way, such that once all deletion mutants are screened for a phenotype, you can be assured that all of the possible deletion mutants have been identified, because you have screened the entire deletion set. The exception is of course essential genes, which are not present in the haploid deletion collection. In addition, a handful of nonessential genes are unable to mate or sporulate, and thus, these genes are not recovered in high-throughput screens using the yeast deletion collection since these two properties are required. Second, if you identify a mutant with an interesting phenotype, you can assign a function to the gene and the identity will be largely certain. In most cases, it is not necessary to clone and confirm the identity of the gene. However, because contamination can occur in the knockout strain collection, you might want to confirm that the strain identified actually bears a knockout allele of the expected gene and does not contain additional mutations contributing to the phenotype that are not associated with the knockout. Because each knockout allele was tagged with a unique "bar code" and flanked with universal primers, identity can simply be confirmed by PCR and a single sequencing reaction (see Experiment XI and Techniques and Protocols 16). Simple genetic analysis such as complementation and tetrad dissection can be used to ensure that the knockout allele is responsible for the phenotype.

The most common use of the deletion mutants is to screen for a particular phenotype (e.g., drug resistance or sensitivity). At the University of Toronto, Charlie Boone has devised a simple way to use the deletion mutants to identify genetic interactions with double mutants (Tong et al. 2001; Tong et al. 2004; Tong and Boone 2007), called synthetic genetic array (SGA) analysis. We will use the SGA scheme (illustrated in Fig. 1) to identify synthetic-lethal interactions for four different mutant alleles (*slx4Δ::natMX, slx1Δ::natMX, rad1Δ::natMX,* and *rtt107Δ::natMX*) when they are combined with the deletion alleles from the knockout collection. The principle of synthetic lethality is that a double mutant is inviable, whereas either single mutant is viable. For example, if you have a deletion of your favorite gene *yfg1Δ* and the strain is viable, and a deletion of an interacting gene (*iag1Δ*) is also viable but the *yfg1Δiag1Δ* double mutant is inviable, then you have a synthetic-lethal interaction. Lethality is the strongest phenotype in this type of screen, but weaker phenotypes—synthetic sickness—where the double mutant has a fitness defect (e.g., slow growth) can also be informative. Before the development of SGA, plasmids were used to identify synthetic-lethal mutants after random mutagenesis with ethylmethane sulfate (EMS). However, we now often use crosses, meiosis, and a powerful haploid selection to identify double mutants and determine synthetic lethality or sickness. A drawback of the original version of SGA (which is what we will use) is that it tested only the nonessential genes and only

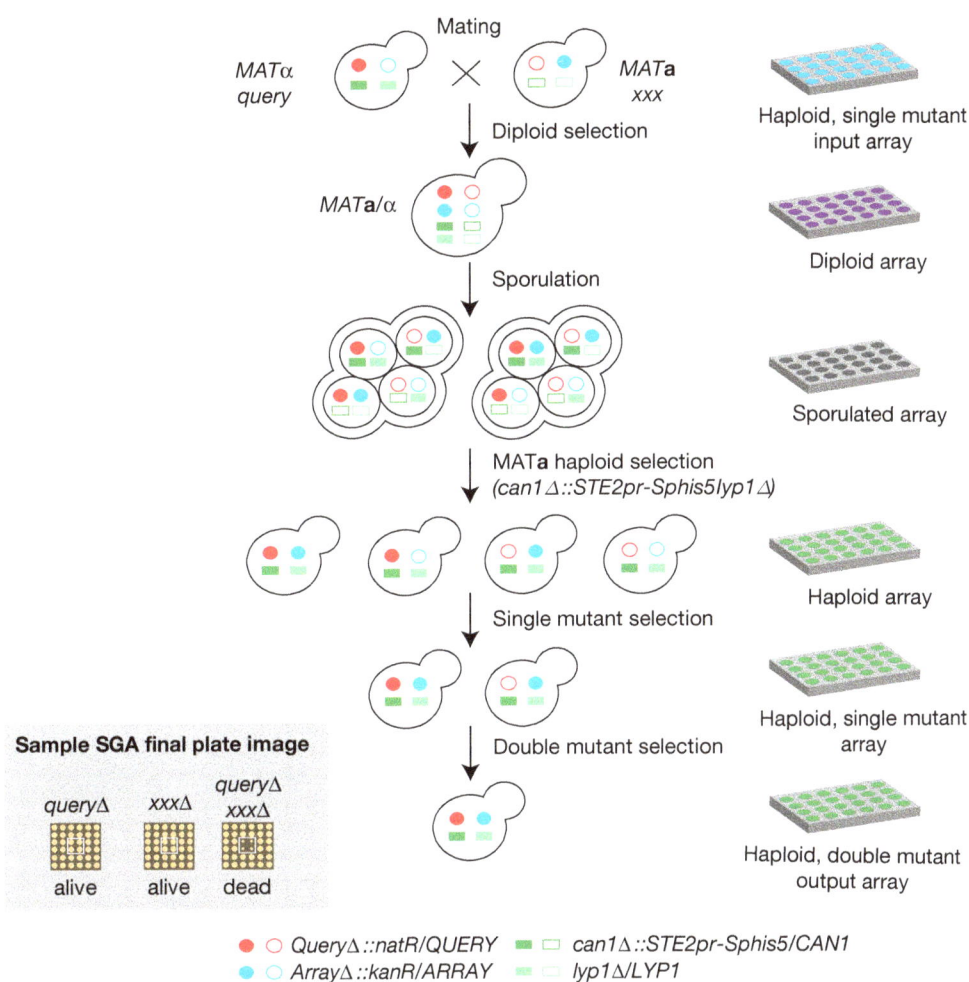

FIGURE 1. Synthetic genetic array (SGA) methodology. A *MATα* query mutant strain carries a deletion in the gene of interest linked to a dominant selectable marker (*natMX*), which confers resistance to the antibiotic nourseothricin (filled red circle). The query strain also carries deletions of the arginine and lysine permease genes, *CAN1* and *LYP1*, respectively. *CAN1* is replaced by the SGA reporter, *STE2pr-his5$^+$*, in which the *STE2 MAT***a**-specific promoter (*STE2pr*) controls the expression of the *Schizosaccharomyces pombe his5$^+$* gene. Deletions of *CAN1* and *LYP1* loci confer sensitivity to canavanine and thialysine, respectively, and are used to select against diploids following the sporulation step. In a typical SGA screen, the query strain is crossed to an ordered array of *MAT***a** nonessential gene deletion strains ("array" mutants) that are marked by a dominant selectable marker, *kanMX*, which confers geneticin resistance (filled blue circle). The resulting heterozygous diploids are transferred by replica pinning to a medium containing reduced carbon and nitrogen sources to induce sporulation. Sporulated cells are then transferred to a synthetic medium lacking histidine but containing canavanine and thialysine to allow for the selective germination of *MAT***a** haploid meiotic progeny. The *MAT***a** haploids are then transferred to a medium containing geneticin to select for array mutants, and then to a medium containing geneticin and nourseothricin to select for double mutants. A sample portion of a plate image following the final SGA selection step is shown in the inset. The deletion of either a query gene (*query*Δ) or an array gene (*xxx*Δ) does not result in any observable fitness defects, but the deletion in both genes is lethal. Shown is an example of an extreme negative genetic interaction termed synthetic lethality. The mutant is represented four times on the array and is highlighted with a white box.

as null alleles. More recently, SGA has been used to screen different kinds of alleles of essential genes (Ben-Aroya et al. 2008; Li et al. 2011), extending the technique to essentially the whole genome.

The innovation behind SGA is a query strain that permits selection of haploids of a single mating type from a complex mixture of sporulating diploids. In typical query strains, the *CAN1* gene has been replaced with a construct containing the *STE2* promoter fused to the *his5+* ORF from fission yeast (*can1Δ::STE2pr-his5⁺*). *his5+* complements *Saccharomyces cerevisiae his3* mutations; but it does not efficiently recombine with the *HIS3* locus. This is important because the *his3Δ1* mutation from the knockout collection strains is not a complete deletion. *STE2*, which encodes the receptor for α-factor mating pheromone, is only transcribed in *MAT**a*** haploids. Thus, the absence of *STE2pr-his5⁺* expression in *MATα* haploids and *MAT**a**/MATα* diploids permits selection of *MAT**a*** haploids on media lacking histidine. The deletion of *CAN1*, which encodes an arginine permease, provides an additional selection for haploids, as *can1Δ* haploids (but not *can1Δ/CAN1* diploids) are resistant to canavanine (a toxic arginine analog). The query strain also has a deletion of the *LYP1* gene, which encodes a lysine permease. The recessive *lyp1Δ* mutation confers resistance to thialysine (a toxic lysine analog), thus providing another selection for haploids.

We have deleted the *SLX4* gene to generate an *slx4Δ::natMX* allele (strain 7-2), deleted the *SLX1* gene to generate an *slx1Δ::natMX* allele (strain 7-3), deleted the *RAD1* gene to generate a *rad1Δ::natMX* allele (strain 7-4), and deleted *RTT107* to generate an *rtt107::natMX* allele (strain 7-5). The nonessential knockout collection has ~5000 strains, each with a different mutation (*orfΔ::kanMX*).

Confirmation of our putative synthetic-lethal interactions is performed by random spore analysis. This is a method to examine meiotic progeny without carrying out tetrad analysis. Random spore analysis is used in this experiment because it allows the analysis of a large number of sporulation products. Tetrad dissection could also be used, but it is of lower throughput. To select for the haploid meiotic progeny, we will use the same method as that used in SGA to eliminate the unsporulated diploid parents and to prevent mating of the haploids. Therefore, haploid single and double mutants are selected on media containing canavanine and thialysine (because the diploids are *CAN1/can1Δ LYP1/lyp1Δ* heterozygotes and thus are sensitive to both drugs), and mating of haploids is prevented by omitting histidine in the media (so that only *MAT**a*** haploids, which express the *STE2pr-his5⁺* marker, are able to grow).

In this experiment, the instructors previously crossed query strains 7-2, 7-3, 7-4, and 7-5 to the complete collection of haploid nonessential gene deletion mutants. We have also included a control cross with a wild-type query strain 7-6, as the scoring routine requires a wild-type cross for normalization. The diploids were previously selected on rich medium (YPD) containing clonNAT and G418, and the diploids were sporulated. The complete procedure, up to pinning to SPO media, has been described in detail by Tong and Boone (2007). You will select the haploids on SD-his-arg-lys medium containing canavanine (can) and thialysine. The result is an array of colonies of haploid *MAT***a** cells that are the meiotic progeny of the heterozygous diploid parents and have the potential to have zero, one, or both deletion mutations. The brilliance of the strain is that haploids are recovered that will not mate among themselves because they are all *MAT***a**. We then test the haploids sequentially for growth on SD/MSG-his-arg-lys + can + thialysine + G418 and SD/MSG-his-arg-lys + can + thialysine + G418 + NAT. If synthetic lethality is present, you cannot recover the double mutant and you will see no growth on G418 + NAT.

Media notes: All SD media are made with the standard formulations except that SD/MSG has 5 mg/mL ammonium sulfate replaced with 1 mg/mL monosodium glutamate as the nitrogen source. Otherwise, G418 is not effective in synthetic medium. Additionally, all the defined SGA media are more like SC than SD! The dropout mixes are based on the SC amino acid mix (not on CSM or other formulations), and this formulation is important for uniform colony growth on high-density arrays. See Appendix A, SGA Media.

The cells can be transferred manually with 384 floating replicator pinning tools purchased from V&P Scientific (http://www.vp-scientific.com/), using rectangular OmniTrays available from Nalgene. In this manual, the cell transfers are accomplished using a "colony copier" template that allows you to orient the transfers with the pinning tool appropriately.

We will use a Singer RoToR colony pinning robot (http://www.singerinstruments.com), which facilitates rapid and reproducible colony pinning, to do our screen. The colonies will be provided on Singer PlusPlates for use with the robot. These are similar to Omnitrays but have an extra notch on one side for alignment in the robot. The colonies are in a 1536 format and 14 plates cover the entire deletion collection. By using the 1536 format, each deletion is present in four copies, and thus, we effectively perform the screen in quadruplicate, which helps eliminate false positives and negatives due to pinning errors.

After completing the selection for haploid double mutants, you will score the arrays using SGATools (Wagih et al. 2013) to identify synthetic genetic interactions. You will then recover the sporulated diploids and retest them by a random spore analysis to validate genetic interactions and to reduce false positives in the screen. Synthetic genetic interactions can also be validated by traditional tetrad analysis. Validation should be done on the same media as the SGA (i.e., do not dissect on YPD to validate an observation made on SD/MSG), and keep in mind that the computational scoring can detect interactions that are not readily discerned by eye (interactions with lower scores can be difficult to validate by visual inspection).

STRAINS

7-1   *MAT**a** orfΔ::kanMX ura3Δ0 his3Δ1 leu2Δ0 met15Δ0* array of deletion mutants, with a border of *MAT**a** ura10Δ::kanMX ura3Δ0 his3Δ1 leu2Δ0 met15Δ0*

7-2   *MATα slx4Δ::natMX can1Δ::STE2pr-his5⁺ lyp1Δ ura3Δ0 his3Δ1 leu2Δ0 met15Δ0*

7-3   *MATα slx1Δ::natMX can1Δ::STE2pr-his5⁺ lyp1Δ ura3Δ0 his3Δ1 leu2Δ0 met15Δ0*

7-4   *MATα rad1Δ::natMX can1Δ::STE2pr-his5⁺ lyp1Δ ura3Δ0 his3Δ1 leu2Δ0 met15Δ0*

7-5   *MATα rtt107Δ::natMX can1Δ::STE2pr-his5⁺ lyp1Δ ura3Δ0 his3Δ1 leu2Δ0 met15Δ0*

7-6   *MATα ura3Δ0::natMX can1Δ::STE2pr-his5⁺ lyp1Δ his3Δ1 leu2Δ0 met15Δ0*

EXPERIMENTAL PROCEDURE

You are provided with 14 PlusPlates containing sporulated diploids. Each pair of students will do one complete screen, either mutant or wild type. Data analysis will be done in groups of four, to combine a mutant screen with a wild-type screen for scoring purposes.

▶ Day 1

Use the Singer RoToR robot to replicate the sporulation plates (SPO) to the first haploid selection plates (HAP1), SD-his-arg-lys + thialysine + canavanine. This selects for *MAT**a*** haploids. Return the sporulation plates to 23°C. Grow the HAP1 plates at 30°C for 2 days.

Notes: *14-plate stacks should be placed in plastic sleeves during incubations and storage to prevent overdrying of the plates. It is important that all array plates be labeled with the query gene, the plate number, and the media type. Check each plate immediately after the replica pinning to make sure that all colonies were transferred accurately. Excess condensation can be wiped from PlusPlate lids using a KimWipe dampened with ethanol.*

▶ Day 3

Use the Singer RoToR robot to replicate the HAP1 plates to media that selects for *MAT**a*** haploids and the array genes (ARR1; SD/MSG-his-arg-lys + thialysine +

canavanine + G418). Store the HAP1 selection plates at 4°C. Grow the ARR1 selection plates at 30°C for 2 days.

▶ Day 5

Use the Singer RoToR to replica-pin the ARR1 plates to the double-mutant selection media (DM1; SD/MSG-his-arg-lys + thialysine + canavanine + G418 + NAT). Store the ARR1 selection plates at 4°C. Grow the DM1 selection plates at 30°C for 2 days.

▶ Day 6 or 7

DM1 plates are scanned after 24–48 h, depending on how healthy the query strain is. Plan on scanning after 24–36 h of growth. Allow plates to warm to room temperature for 30 min before scanning. We will score for fitness defects by using SGATools (http://sgatools2.ccbr.utoronto.ca) to measure colony sizes and compare them to the wild-type control screens. Read the SGATools paper (http://nar.oxfordjournals.org/content/41/W1/W591.full.pdf).

▶ Day 7 and 8

Use SGATools to identify synthetic genetic interactions, as detailed in Techniques and Protocols 15. Generate a list, in rank order, of synthetic genetic interactions. Each group of four will generate a list of SGA scores for one mutant query (*SLX1*, *SLX4*, *RTT107*, *RAD1*). The top negative genetic interactions from these lists will be validated in random spore analysis by each group of four.

▶ Day 11

We will perform random score analysis (RSA) for the top 20 negative genetic interactions from each screen, so each student will perform five RSA assays.

Remove the SPO plates from the refrigerator. Add 1 mL of sterile H_2O to each of five Eppendorf tubes. For each analysis, pick *one* colony of the relevant sporulated diploid (e.g., if you want to analyze *slx4*Δ × *sgs1*Δ, then find the correct coordinates for *SGS1* on the array [plate, column, row]), find the four *SGS1* replicates at those coordinates on the SPO plate from the *SLX4* SGA screen, and pick one of those replicates). Resuspend the colony of spores in 1 mL of dH_2O. Place parafilm on the sporulation plates and return them to the refrigerator. Vortex the cells for ~30 sec to disperse them in H_2O. Plate the cells as follows:

- 20 μL on SD-his-arg-lys + thialysine + canavanine

- 40 μL on SD/MSG-his-arg-lys + thialysine + canavanine + G418

- 40 μL on SD/MSG-his-arg-lys + thialysine + canavanine + NAT

- 80 μL on SD/MSG-his-arg-lys + thialysine + canavanine + G418 + NAT

Incubate plates at 30°C.

▶ Day 12

Score the RSA plates. Double mutants with synthetic genetic interactions will fail to grow, or they will produce smaller colonies on the DM selection plate (SD/MSG-his-arg-lys + thialysine + canavanine + G418 + NAT) when compared to the HAP (SD/MSG-his-arg-lys + thialysine + canavanine + G418), ARR (SD/MSG-his-arg-lys + thialysine + canavanine + G418), or QUERY (SD/MSG-his-arg-lys + thialysine + canavanine + NAT) selection plates. Score your interactions (+++ = no growth, ++ = very small colonies, + = detectably small colonies, − = same colony size as HAP plate) and scan your plates.

▶ Day 14 and 15

Prepare your results for presentation to the class. Use the following as guidelines:

- Using SGD, PubMed, and other resources, what is known about the function of Slx4, Slx1, Rtt107, and Rad1? Are there data that might be relevant to understanding the genetic interactions that you will describe to us?

- What synthetic-lethal and synthetic-sick interactions did you observe?

- How do these compare with genetic interactions that have been reported (Bio-Grid [http://thebiogrid.org]; DryGIN [http://drygin.ccbr.utoronto.ca])?

- Can all of the hits that you detected with SGATools be verified by random sporulation? Is there any relationship between SGA score and RSA score?

- SGATools discards some genes that are true positives. What are these genes, and why are they discarded?

- Is there any enrichment for GO terms that might shed light on the function of the query genes? (You can use tools and links on SGD and BioGRID, or GO term finder [http://go.princeton.edu/cgi-bin/GOTermFinder], for this analysis.)

- Plot the overlap between the different pairs of screens (use BioVenn [http://www.cmbi.ru.nl/cdd/biovenn/]). Does the overlap suggest anything about function?

- What are your conclusions/thoughts about this methodology? Did you learn anything about SLX1/SLX4/RTT107/RAD1?

MATERIALS

Supplies for Singer RotoR Robot

Singer PlusPlates, PLU-001
Repads 384L, RP-MP-3L
Repads 1536, REP-005

Pre-course

YPD PlusPlates
YPAD + G418 PlusPlates
YPD + G148 + clonNAT PlusPlates
SPO PlusPlates
384 Long-pin Repads
1536 Repads
*MAT***a** yeast deletion collection

Day 1	SD-his-arg-lys + thialysine + canavanine PlusPlates 1536 Repads
Day 3	SD/MSG-his-arg-lys + thialysine + canavanine + G418 PlusPlates 1536 Repads
Day 5	SD/MSG-his-arg-lys + thialysine + canavanine + G418 + NAT PlusPlates 1536 Repads
Day 6 or 7	Scanner Thumb drive
Day 11	Sterile H_2O 10 Eppendorf tubes 10 SD/MSG-his-arg-lys + thialysine + canavanine + G418 + NAT plates 10 SD/MSG-his-arg-lys + thialysine + canavanine + G418 plates 10 SD/MSG-his-arg-lys + thialysine + canavanine + NAT plates 10 SD-his-arg-lys + thialysine + canavanine plates sterile glass beads
Day 12	Scanner Thumb drive

REFERENCES

Ben-Aroya S, Coombes C, Kwok T, O'Donnell KA, Boeke JD, Hieter P. 2008. Toward a comprehensive temperature-sensitive mutant repository of the essential genes of *Saccharomyces cerevisiae*. *Mol Cell* **30:** 248–258.

Giaever G, Chu AM, Ni L, Connelly C, Riles L, Veronneau S, Dow S, Lucau-Danila A, Anderson K, Andre B, et al. 2002. Functional profiling of the *Saccharomyces cerevisiae* genome. *Nature* **418:** 387–391.

Li Z, Vizeacoumar FJ, Bahr S, Li J, Warringer J, Vizeacoumar FS, Min R, Vandersluis B, Bellay J, Devit M, et al. 2011. Systematic exploration of essential yeast gene function with temperature-sensitive mutants. *Nat Biotechnol* **29:** 361–367.

Tong A, Boone C. 2007. High-throughput strain construction and systematic synthetic lethal screening in *Saccharomyces cerevisiae*. *Methods Microbiol* **36:** 369–386.

Tong AH, Evangelista M, Parsons AB, Xu H, Bader GD, Page N, Robinson M, Raghibizadeh S, Hogue CW, Bussey H, et al. 2001. Systematic genetic analysis with ordered arrays of yeast deletion mutants. *Science* **294:** 2364–2368.

Tong AH, Lesage G, Bader GD, Ding H, Xu H, Xin X, Young J, Berriz GF, Brost RL, Chang M, et al. 2004. Global mapping of the yeast genetic interaction network. *Science* **303:** 808–813.

Wagih O, Usaj M, Baryshnikova A, VanderSluis B, Kuzmin E, Costanzo M, Myers CL, Andrews BJ, Boone CM, Parts L. 2013. SGATools: One-stop analysis and visualization of array-based genetic interaction screens. *Nucleic Acids Res* **41:** W591–W596.

Measuring Mutation Rates and Studying Human Genetic Variation in Yeast

As famously proved by Luria and Delbrück (1943), the occurrence of mutations is a stochastic process. Throughout the growth of a population of cells, individuals acquire mutations. The equation below describes continuous exponential growth:

$$N(t) = N_0 e^{kt}$$

where $N(t)$ is the number of cells at time t, N_0 is the initial number of cells, k is the growth rate, and t is time. In addition, define m as the number of mutation events and μ as the number of mutation events per cell division or the mutation rate:

$$\mu = \frac{m}{(N(t) - N_0)}$$

Because we can count the number of cells at the beginning and the end of the culture, we can get the number of cell divisions. In addition, $N(t)$ is generally $\gg N_0$, and thus, we generally neglect N_0. m is a little harder.

Unfortunately, we cannot directly measure m. We can only measure the number of mutants, not the number of mutation events. (Note that m does not necessarily refer to events that are all identical molecularly. For example, many different base pair changes could result in the drug resistance that we will observe.) The timing of the mutation has a huge effect on the number of mutants in a culture started from a small number of cells. For example, Cultures 1 and 2 below both have experienced one mutation ($m = 1$) over seven cell divisions, but Culture 2 has a much higher proportion of mutant cells in the last generation (4/8 vs. 1/8). Culture 3 has the same proportion of mutant cells as Culture 2 (4/8), but more mutation events (3 vs. 1).

Simply assuming a normal distribution and taking the mean of many cultures would not accurately reflect the mutation rate. The variation from culture to culture

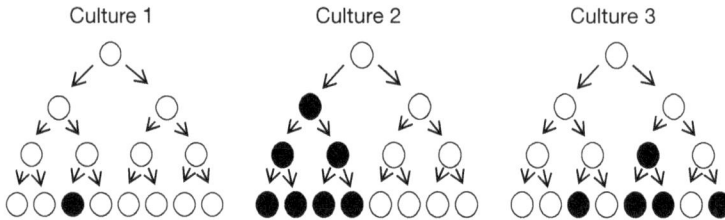

is huge (as you will measure), and the mean is consequently unstable. Instead, mutations are best described by a Poisson distribution. Because a culture with one early mutation could have the same fraction of mutant cells as a culture with two later mutations, it is impossible to classify the cultures according to the number of mutational events. Cultures containing zero mutants (and one mutant, although this is more complicated to use) are unambiguous. From many independent cultures, we can measure the proportion of cultures containing zero mutations and use the Poisson distribution to generate a good estimate of the mutation rate. This type of experiment is called a fluctuation test. Luria and Delbrück first performed this test with bacteria, and 70 years later, it is still one of the most commonly used and most accurate ways to measure mutation rate.

Wild-type yeast are sensitive to the drug canavanine, a toxic analog of the amino acid arginine. The drug is imported into the cell via the transporter Can1. When the *CAN1* gene is mutated to inactivity, the cells become resistant to canavanine because they can no longer transport it into the cell. We will measure the mutation rate from canavanine sensitivity to canavanine resistance in two different yeast strains: a wild-type strain and a strain with one of its DNA repair genes, *MSH2*, deleted. Most mutations that occur in a cell are detected and fixed by a host of repair proteins. When these proteins are disabled, the mutation rate of the cell increases.

These genes are of more general interest because they tend to be disabled in human cancers. In particular, inherited mutations in the human version of *MSH2* predispose carriers to colon cancer. Because *MSH2* is highly conserved, mutations can be created in the yeast protein in the same residue as in the human protein. Alison Gammie's lab has pioneered the use of yeast as a platform for testing the function of human genetic variation, specifically mutations found in patients with hereditary colon cancer (Gammie et al. 2007). She discovered that cancer-causing mutations lead to a loss of function phenotype (e.g., increased mutation rate), whereas other mutations do not. These mutations may be noncausative, or they may interfere with a process that is not well modeled by yeast, such as a particular protein–protein interaction or a splicing problem. Each group will test a strain that carries one potentially cancer-causing mutation recreated in yeast by the Gammie lab. You will have to make a prediction about whether the allele would in fact raise disease risk in a patient. These alleles have been tested qualitatively, but not by the fluctuation test.

We can get a qualitative view of the mutation rate by using a different pheno-type of the strains. These strains carry a mutation in the gene *ADE2*. The Ade2 pro-tein (aka phosphoribosylamino-imidazole-carboxylase) performs an enzymatic step in the synthesis of the nucleotide adenine. The chemical reaction performed by Ade2 does not occur in an *ade2* strain, causing the intermediate, 5′-phosphoribosyl-5-aminoimidazole, to accumulate in the cell. This buildup causes the cells to turn red. Cells carrying a normal copy of *ADE2* contain very little of the intermediate because it is quickly metabolized by the functional enzyme, and they are cream-colored. We have added just enough adenine to the plates so the cells can survive despite the *ade2* mutation, yet not enough adenine to prevent the accumulation of the red pigment.

The particular mutant allele carried by these strains is a nonsense mutation. On occasion, the cells "revert" to the cream-colored, *ADE2* state, either by mutating the nonsense codon or by mutating another gene that can suppress the nonsense codon. To get a qualitative estimate of the mutation rate, you can look at the fre-quency of the occasional cream cells that are present in a sea of red, especially on the densely streaked area of the plate. The frequency at which revertants appear can give you an estimate of the mutation rate. However, this may also reflect the rate at which such mutants were already present in the overnight culture that was frozen so it can be misleading.

STRANS

8-1	*MAT**a** msh2Δ::URA3 CAN1 hom3-10 ade2-1 trp1-1 ura3-1 leu2-3,112 his3-11,15, pRS413 (CEN/ARS HIS3)*
8-2	*MAT**a** msh2Δ::URA3 CAN1 hom3-10 ade2-1 trp1-1 ura3-1 leu2-3,112 his3-11,15, pMSH2 (MSH2 CEN/ARS HIS3)*
8-3 through 8-10	*MAT**a** msh2Δ::URA3 CAN1 hom3-10 ade2-1 trp1-1 ura3-1 leu2-3,112 his3-11,15, pMSH2-allele (msh2-allele CEN/ARS HIS3)*

EXPERIMENTAL PROCEDURE

We will use a modification of the fluctuation test by Greg Lang, Andrew Murray (Lang and Murray 2008), and Alison Gammie that improves the throughput of the assay.

▶ Day 1

The instructors will streak wild type (wt), *msh2*, and the mystery allele strains from the freezer to His dropout plates and place them in the incubator. Each group will get a different mystery allele to characterize.

▶ Day 3

For each strain, inoculate 5 mL of SC-His with a single colony and place in roller drum overnight.

Allow the plate to continue to overgrow. This will allow the adenine to run out so you can look at the rate of Ade + revertants and estimate the mutation rate.

▶ Day 4

You will grow the independent cultures in round-bottom 96-well plates. You will need four plates: one for the wt, one for the *msh2* control strain, and two plates for the mystery allele. In each plate, fill the outer ring of wells with 100 μL of SC-His medium to alert you to possible contamination.

Each culture will be inoculated with 1000 cells. Because 1000 is much less than the frequency of the mutant in the starting culture, you can safely assume that all 1000 of the cells are canavanine-sensitive.

We have already determined the target population size for cultures with high and low mutation rates. For the mystery alleles, you do not yet know what to expect, so make one plate at each volume. Count the cell concentration in the overnight culture using the hemocytometer (Techniques and Protocols 12) and/or the C6 (Techniques and Protocols 3). Determine what dilution of the overnight culture you will need to get 1000 cells/100 μL (i.e., 10^4 cells/mL) for the wt and mystery allele plates. Dilute to a final concentration of 1000 cells/20 μL (5×10^4 cells/mL) for the *msh2* and the second plate of your mystery allele.

Make 10 mL of each appropriate dilution using SC–His medium. Right before use, mix thoroughly and pour into the pipetting reservoir.

Label each 96-well plate on both the lid and the side. Use a multichannel pipettor to pipette the appropriate volume of solution (100 μL for the wt and mystery allele, and 20 μL for the *msh2* and mystery allele) into the central 60 wells of the 96-well plates. Remove tips from the tip box so that you use six each time, avoiding the media-filled edge wells. Be as accurate as possible because more or less media will alter the number of yeast cells you have in the end, and the population size determines the probability of getting zero canavanine-resistant colonies in a culture.

Seal the plates well with foil tape, placing the plastic lid back on top once sealed. Place them in the 30°C incubator (not shaking) for 2 d.

▶ *Day 5*

Check the 96-well plates to make sure cells are growing. Look at the overgrown cell streaks. Notice the sectoring pattern and how it relates to mutation rate. Try to predict if your mystery allele is more like wt or the knockout allele.

▶ *Day 6*

Cell counting to get $N(t)$: To calculate the mutation rate, you need to know $N(t)$, the current number of cells. You will count the cells in a hemocytometer (see below and Techniques and Protocols 12) and with the C6 cytometer (Techniques and Protocols 3). To get an idea of how reproducible the replicate cultures are, you will count multiple cultures for each strain.

1. Centrifuge the 96-well plates briefly in the swinging bucket centrifuge to remove condensation from the foil. Remove the foil from each plate.

2. For the two 96-well plates with just 20 µL, add 80 µL of H_2O to bring them to 100 µL. To avoid cross-contamination, change tips between wells.

3. Vigorously pipette one well up and down several times until the culture is uniformly cloudy. Make sure to use a new pipette tip for each well. The cells will settle again if you let them sit for too long, so resuspend each well right before you use it.

4. Pipette enough culture to fill a counting chamber.

5. View the cells under the 20× or 40× objective, adjusting the light intensity until you can see the cells clearly. If the cells are clumpy, you might need to

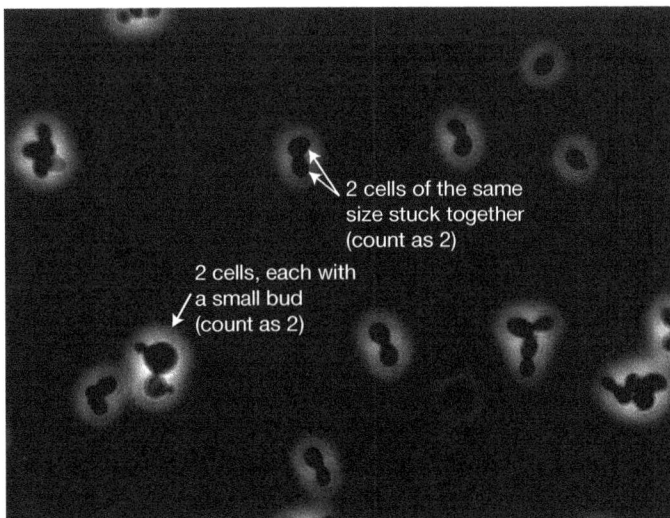

2 cells of the same size stuck together (count as 2)

2 cells, each with a small bud (count as 2)

resuspend them better and make a new slide. If there are too many of them, make a dilution and a new slide. Count enough fields to get at least 100 cells. Do not count buds, which you can recognize as smaller than the mother cells.

6. Repeat until you have counted three cultures of each strain.

7. Determine the average number of cells/well for each strain. Knowing that the cultures each started with 1000 cells, how many generations have they grown?

8. Count an additional three wells using the C6. Did you get the same number? If you are off by orders of magnitude, recheck your technique and calculations.

Plating for Mutants

For each strain, you should plate 54 independent cultures of cells. To accurately use the Poisson distribution, a certain fraction of the plates should contain zero mutants. We have already figured out the amount of cells to use for each strain. If you were starting this experiment from scratch, you would have to try several different population sizes until you found the amount that gave an adequate result.

1. Because you will be plating the entire volume of each well, the canavanine plates must be very dry. To this end, we will absorb moisture from the plates using sterile Whatman filter paper. Cover the replica-plating block with a sterile velvet. Use flamed forceps to place a sterile circle of Whatman filter paper on top of the velvet. Press the canavanine plate onto the paper to transfer it to the surface of the plate. Repeat for 24 canavanine plates. Allow them to sit at room temperature for 30 min.

2. Use sterile forceps to remove the filter paper from each plate just before use. The dried plate will accommodate nine carefully plated cultures. Use the template below to help place them.

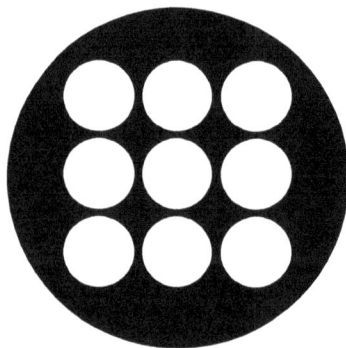

3. Pipette up and down to completely resuspend the cells in one well.

4. Pipette up the *entire volume* of the well and deposit it onto the canavanine plate in one of the indicated spaces. Change tips for each well. Be careful not to allow adjacent spots to run into each other while pipetting: Handle the plates gently and keep them level. Allow the plates to sit undisturbed until all of the liquid is absorbed and then transfer them to the incubator.

▶ *Day 9*

1. Examine the colonies on each patch. Each colony is the result of a single canavanine-resistant cell present in the plated culture. However, we do not know if multiple colonies in a given patch derived from independent mutation events or growth of a single parent.

2. The only time the number of mutations matches the number of mutants is for the zero class. Find the proportion of patches with no mutants, P_0. It should hopefully be ~10%–80%, which is the accurate range for this assay.

3. According to the Poisson distribution, $P_n = e^{-m}m^n/n!$, with n being the number of mutations and m the average number of mutations. Therefore, $P_0 = e^{-m}$. The m is the same m as in the mutation rate equation. From your cell counts on Day 6, you have measured $N(t)$. From the canavanine plates, you have measured P_0. Now you can calculate the mutation rate:

$$\mu = \frac{-\ln(P_0)}{N}.$$

4. The units should be mutations per locus per cell division. By how much did the *msh2* mutation elevate the mutation rate at the *CAN1* locus? How does your mystery allele compare with the controls?

5. Tell us the proportion of patches with no mutants for each strain. We will collect the class data set for you to analyze along with your own results.

MATERIALS

Note: Amounts provided are the requirements for each pair of students.

General   SC-His liquid media
             Basins for multichannel pipettes

Day 1   1 SC-His plate

Day 3   3 Tubes of 5 mL of SC-His medium

Day 4   4 U-bottom 96-well plates
                     4 Foil seals

Day 6   24 SC-His-Arg + canavanine plates
                     24 Sterile Whatman paper circles

REFERENCES

Gammie AE, Erdeniz N, Beaver J, Devlin B, Nanji A, Rose MD. 2007. Functional characterization of pathogenic human MSH2 missense mutations in *Saccharomyces cerevisiae*. *Genetics* **177**: 707–721.

Lang GI, Murray AW. 2008. Estimating the per-base-pair mutation rate in the yeast *Saccharomyces cerevisiae*. *Genetics* **178**: 67–82.

Luria S, Delbrück M. 1943. Mutations of bacteria from virus sensitivity to virus resistance. *Genetics* **28**: 491–511.

Mutation Detection Using Comparative Genomic Hybridization

Mutations are the heart of genetics. The goal of a classic genetic screen is to isolate strains with phenotypes of interest caused by mutations, either spontaneous or induced by a mutagen. Most of the time we are interested in determining the molecular nature of the mutation of interest. In Experiments IX and X, you will use genomic methods to characterize genomes, discover mutations, and link these mutations to phenotype.

DETECTING COPY-NUMBER CHANGES BY ARRAY COMPARATIVE GENOMIC HYBRIDIZATION

Changes in gene and chromosome copy number fall into the class of mutations known as copy-number variations, or CNVs. Yeast are amazingly plastic with respect to the frequency and extent of CNV they can tolerate. Although most large-scale CNVs such as whole-chromosome disomy are generally deleterious, some CNVs can cause increases in cell fitness under particular circumstances. For example, even very soon after the generation of the yeast deletion collection, an estimated 8% of strains were already aneuploid (Hughes et al. 2000). These segmental amplifications frequently contained a paralog of the deleted gene, and the amplification resulted in phenotypic suppression. Many other examples can be found in industrial strains, experimental evolution, and fungal pathogenesis.

We will use array-based comparative genomic hybridization (CGH) to discover CNVs. Although whole-genome sequencing can also discover CNVs, microarrays are still a handy and often easier method for screening. We will measure relative copy number between two genomes by isolating DNA from a wild-type (wt) strain and a derived mutant, differentially labeling the DNA with fluorescent nucleotides, and competitively hybridizing the two DNAs to a microarray containing a single 60-mer probe specific to each gene. Other available arrays can give much higher resolution, including tiling microarrays, which contain overlapping 25-mer probes every four bases. Typically, this level of resolution is not required, because the majority of CNVs have stereotypical breakpoints at repeat elements such as Ty retrotransposons. This knowledge makes it reasonably accurate to estimate the breakpoints for most events. It is important to realize, however, that arrays only provide the copy number of a particular DNA sequence; they do not provide information about the structure of the CNV or the breakpoint junctions. To learn about the structure, further analysis is required. Pulsed-field gel electrophoresis (PFGE) allows whole chromosomes to be visualized on a gel. Polymerase chain reaction (PCR) across proposed novel breakpoints can confirm hypothesized junctions. Many other molecular biology techniques (Southern blots, restriction mapping, etc.) can also be brought to bear on further characterization.

Another ambiguity in CGH is the basal copy number. CGH can only tell the relative copy number of sequences: A diploid strain with a segment at four copies will appear to be identical to a haploid strain with a duplication. Again, other methods are required to determine this variable. One effective technique that we will use in this lab is flow cytometry to measure DNA content compared with known controls.

FINDING MUTATIONS AND LINKING THEM TO PHENOTYPE BY SEQUENCING

Of course, sequencing can also reveal CNVs. The number of times a given sequence is sampled is directly proportional to the copy number. Next-generation sequencing allows high coverage of a genomic sample, leading to very accurate estimates of DNA copy number. However, the typically short sequence reads cannot bridge long repetitive elements, making most next-generation technologies of little use for breakpoint and structural characterization. Paired end sequencing, where two ends of a clone of defined length are sequenced in such a way as to link them, is somewhat better in this respect, although the currently practical sizes are still much smaller than a Ty element. However, short repeat sequences are amenable to this approach.

Next-generation sequencing is very effective at discovering single-nucleotide variants (SNVs). A sequence obtained from a strain of interest is mapped back against a reference genome, and differences between the two are scored on the basis of quality and coverage. SNVs can consequently be discovered with a low false-positive rate and high sensitivity. Heterozygous SNVs can be more difficult to call than haploid or homozygous SNVs, but they are straightforward given enough sequence coverage (typically > 30×, which means that each base is covered by an average of 30 reads).

Small insertions and deletions ("indels") are more difficult to discover and require extra attention in the analysis. Reads that map to multiple locations are also typically excluded from analysis, making as much as 10% of the genome off limits for mutation discovery. Depending on the exact sequencer technology used, homopolymer runs of a single base can also pose problems.

Tiling microarrays can also detect SNVs and indels. This technology is based on detecting a decrease of hybridization of the short oligonucleotide probes to a mutant sequence that harbors mismatches. Most, although not all, tiling array methods highlight the location of the mutation but are unable to determine the exact nature of the base change. Typically, further validation via Sanger sequencing is required. Sanger verification is also frequently performed to confirm mutation calls from next-generation sequencing as well, given the relatively high error rate of these platforms.

FOLLOW-UP OPTIONS ONCE YOU HAVE CANDIDATE MUTATIONS

Just because a mutation is present does not mean that it is causative for phenotype. When using a mutagen, for example, many mutations may be present in a genome. Even without mutagens, mutations accumulate randomly during growth. These mutations would then be fixed in any clonal derivatives. For these reasons, proper strain hygiene is important to keep genetic backgrounds clean. Avoid unnecessary propagation and extended storage of strains on plates. Pay close attention to crosses to ensure only isogenic lines are crossed. Even strains of the same background from different labs can carry a surprising number of polymorphisms relative to one another.

One classic method of proving that a mutation is related to a phenotype is linkage analysis. We will be going through the basics of tetrad analysis in Experiment IV. Segregants from a backcross would simply be genotyped for each of the candidate mutations, and the causative mutation would show linkage to the phenotype. Linkage can be integrated with whole-genome sequencing in a single step to efficiently identify and screen candidate mutations (Birkeland et al. 2010). This method, sometimes called bulk segregant analysis, is what we will use in Experiment X.

The strain with the phenotype of interest is backcrossed to an otherwise isogenic wt strain. The diploid is sporulated, and the tetrads are typed for segregation of the trait of interest. In the simplest case where the phenotype is caused by a mutation in a single locus, the trait will segregate 2:2. All segregants with the trait are pooled, and all without the trait are combined into a second pool or "bulk." The two pools are then sequenced. The causative mutation will be found in all genomes of the first pool and none in the second. Unlinked mutations will be found at 50% allele frequency in both pools. Linked mutations will show a bias depending on their distance from the causative mutation.

Many other methods are used for following up candidate mutations. The mutation can be mapped by linkage via crossing to the entire deletion collection (Jorgensen et al. 2002). Although this sounds daunting, most of the manipulations can be accomplished by pinning and is even faster with robotic automation. If the mutation is recessive, it can be complemented with a clone carrying the wt sequence. Several libraries are available that contain individual genes or short genomic segments tiling over the entire genome (e.g., Ho et al. 2009; Hvorecny and Prelich 2010). Conversely, the mutant strain can be crossed to the corresponding deletion strain, in which case, lack of complementation would indicate the causative mutation.

Finally, the individual mutation could be reconstructed via site-directed mutagenesis and gene replacement in a fresh strain as a definitive test of causation.

STRAINS

9-1 through 9-8	Mystery strains
9-9 through 9-16	wt strains matched to the mystery strains
9-17	Haploid reference strain
9-18	Diploid reference strain
9-19	Triploid reference strain
9-20	Tetraploid reference strain

EXPERIMENTAL PROCEDURE

We have already prepared DNA from a selection of strains. Each group will receive a tube of wt DNA and mystery DNA. We have made the DNA by the modified Hoffman–Winston prep you will use in Labs X and X1 (Techniques and Protocols 4). It is important to use the same protocol for the wt and mutant DNA preps, because different preps appear to differentially purify subtelomeric DNA.

You will fragment the DNA, label the experimental and reference sample with different fluorescent dyes, and clean up the samples. We will then perform the array hybridization for you and you will analyze the data to determine the genome content of your mystery strain. You will also use flow cytometry to compare the DNA content of your mystery strain with that of control strains of known ploidy.

▶ *Day 1*

Measure DNA concentration using the Qubit fluorometer (see Techniques and Protocols 16). Digest 4 µg of wt and query strain DNA with *Hae*III:

Genomic DNA	— µL
10× NEBuffer 4	4 µL
*Hae*III	2 µL
H_2O	to 40 µL

Incubate your two reactions at 37°C for 1 h.

Carry out the Zymo column cleanup (follow the kit's instructions) and elute with 25 µL of H$_2$O. Use the Nanodrop and 1.5 µL of sample to determine yield. Freeze your DNA.

▶ Day 2

Proceed with the Bioprime labeling kit as per its instructions. Make sure to use different dyes for the two samples and record which dye is which. Clean up the reaction using the Zymo columns as directed except using 500 µL of binding buffer. Elute in 25 µL H$_2$O.

Use the Nanodrop to measure DNA yield and dye incorporation, using the preset wavelengths appropriate for Cy3 and Cy5. Yield should be ~25 pmol dye/µg.

The instructors will hybridize the arrays. For reference, the following instructions are provided for hybridization, washing, and scanning the array.

▶ Day 3 (Performed by Instructors)

Add 1250 µL H$_2$O to a tube of 10× Agilent blocking agent. Vortex and incubate at 37°C until the pellet is resuspended. We are using slides with eight arrays per piece of glass. Mix up the hybe cocktail:

Cy3-labeled DNA	100 ng
Cy5-labeled DNA	100 ng
H$_2$O	to 20 µL
10× Agilent blocking agent	5 µL

Heat at 95°C for 5 min. Allow to cool briefly to room temperature.

Add 25 µL of 2× Hi-RPM hybridization buffer. Mix by gentle pipetting.

Centrifuge for 1min at 13,000 rpm.

Load 40 µL of the 50 µL onto the array, avoiding any debris at the bottom of the tube.

Place a gasket slide, Agilent side up, in a hybe chamber.

Pipette the appropriate volume of probe, avoiding bubbles, onto the center of one gasket area. Do not eject the last microliter or two in order to avoid bubbles. Spread it around as you pipette, but not too close to the gasket.

Do the same for the other gasket area with the next probe.

Remove the array from the box. The Agilent side is the array side, and thus, should face down, onto the probe.

Carefully lower the array over the gasket slide, keeping it level.

Once the array is resting on the gasket slide, put the top of the hybe chamber on, and slide on the screw assembly. Tighten the screw all the way down, tight.

Look through the back of the chamber and rotate the slide. There should be one big bubble that moves freely. There may be one big bubble and a couple of little ones stuck to the sides. If they are small and isolated, do not worry too much about them. You will probably do more harm than good trying to remove them. If they seem like they will interfere with the array, you can try knocking the array with the heel of your hand to dislodge them.

Place the array in the hybridization oven, making sure to balance the rotisserie. Hybridize at 65°C, 20 rpm for 24 hours.

▶ Day 4

Prepare wash solutions. Rinse the wash chambers, racks, and stirbars with deionized H_2O.

Setup:

One Wash-A chamber for disassembling the sandwich (a small jar or beaker works—just make sure that the sandwich can be submerged).
One Wash-A chamber with a rack and a stirbar on a stirplate.
One Wash-B chamber with a stirbar on a stirplate.
One acetonitrile chamber with a stirbar on a stirplate.
One optional chamber with Agilent's stabilization and drying solution, if the ozone is above 30 ppb, which it will be on Long Island in the summer.

All of the washes should be stirring so that they are visibly turbulent. Make sure that the entire slide is submerged at all times.

Disassemble each hybe chamber one at a time using the plastic tweezers to gently wedge open the sandwich while submerged in Wash A. Transfer slide to the rack in the other Wash-A chamber. Leave a gap between each slide and between the slides and the wall.

Once all the slides are in the rack, stir for 1 min.

Transfer the rack into Wash B and stir for exactly 1 min. Do not worry about transferring some Wash A into Wash B.

Quickly transfer the rack into the acetonitrile, draining off some of the Wash B as you go.

Let stir 30 sec (only 10 sec if using stabilization and drying solution, then quickly transfer to stabilization and drying solution, and stir for 30 sec).

Slowly and evenly pull the rack out of the final wash chamber (acetonitrile, or acetonitrile-based stabilization and drying solution). If you see droplets remaining on the slides, submerge them, and try again.

Set the rack on a KimWipe.

Scan the array using the Agilent scanner.

We will return a quality control report and a file containing the data collected from the scanned image.

▶ Day 5/6

Data analysis: Extract the gene name and log ratio columns from the data file using Excel or another program of choice. Make sure that the file format matches the example pcl format given. The log ratio column has already been normalized and processed. Normalization is required to account for unequal labeling, hybridization, and signal capture for the different samples and dyes.

> Open the pcl file in Java Treeview
> Select the "karyoscope" option
> Select the Agilent yeast coordinate file

Look for aberrant segments. What types of structures are these consistent with? Can you determine what copy number each region in the genome is at? Are multiple configurations consistent with the data? Prepare a hypothesis about the structure of any rearrangements and the sequences at the breakpoints.

▶ Day 9

Inoculate overnight cultures of your mystery strain, reference strain, 9-17, 9-18, 9-19, and 9-20 in 2.5 mL of YPD.

▶ Day 10

In the morning, dilute the cultures back to an OD of 0.05 so that they will be actively dividing when you harvest them later in the day.

Prepare the cells for flow cytometry, staining the cells with SYBR Green to determine DNA contents (follow Techniques and Protocols 13 to prepare and stain the cells, and Techniques and Protocols 3 to run the cytometer).

How does your mystery strain compare to the controls? Does this resolve any of the ambiguities from your initial analysis of your CGH data?

MATERIALS

Note: *Amounts provided are the requirements for each pair of students.*

Day 1   *Hae*III

DNA Clean and Concentrator-5 columns (D4004 Zymo Research)

Day 2 Bioprime Array CGH Genomic Labeling module (Invitrogen/Life Technologies 18095-012)
Cy3 dUTP and Cy5 dUTP (NEL578001EA and NEL579001EA NEN/Perkin Elmer)

Day 3 Hybridization reagents (Agilent)

Day 10 50 mм sodium citrate (pH 7.4)
RNase A
Proteinase K
SYBR Green
2-mL tubes for C6

REFERENCES

Birkeland SR, Jin N, Ozdemir AC, Lyons RH, Weisman LS, Wilson TE. 2010. Discovery of mutations in *Saccharomyces cerevisiae* by pooled linkage analysis and whole-genome sequencing. *Genetics* **186:** 1127–1137.

Ho CH, Magtanong L, Barker SL, Gresham D, Nishimura S, Natarajan P, Koh JL, Porter J, Gray CA, Andersen RJ, et al. 2009. A molecular barcoded yeast ORF library enables mode-of-action analysis of bioactive compounds. *Nat Biotechnol* **27:** 369–377.

Hughes TR, Roberts CJ, Dai H, Jones AR, Meyer MR, Slade D, Burchard J, Dow S, Ward TR, Kidd MJ, et al. 2000. Widespread aneuploidy revealed by DNA microarray expression profiling. *Nat Genet* **25:** 333–337.

Hvorecny KL, Prelich G. 2010. A systematic CEN library of the *Saccharomyces cerevisiae* genome. *Yeast* **27:** 861–865.

Jorgensen P, Nelson B, Robinson M, Chen Y, Andrews B, Tyers M, Boone C. 2002. High-resolution genetic mapping with ordered arrays of *Saccharomyces cerevisiae* deletion mutants. *Genetics* **162:** 1091–1099.

EXPERIMENT X

Mutation Detection Using Whole-Genome Sequencing and Linkage

We will use pooled genome sequencing to identify mutations that emerge during the course of a laboratory evolution experiment and link them to a particular "evolved" phenotype.

Yeast of the S288c background typically grow as well-dispersed single cells that do not form clumps or a biofilm on glass surfaces. In a series of laboratory evolution studies, this strain has been propagated for hundreds of generations in a continuous culture device called a chemostat. Because of the constant flow of media into and out of the chemostat, there was a strong selection for mutants that are able to stick to the walls or form large clumps that sink to the bottom of the culture tube, thus increasing the residence time and competitive fitness for these mutants. Because clump formation clogs the device and limits the duration of evolution experiments, it would be valuable to identify the mutations that cause it so that ancestral strains can be engineered that no longer evolve the trait.

Several of these evolved strains were isolated and backcrossed to the ancestral strain of the opposite mating type. You will study two of the mutants in which this phenotype segregates as a single locus in the backcross. The individual segregants have been frozen in 96-well plates. You will pin the strains to YPD, type the segregants to confirm the segregation pattern, pool them, and sequence them.

For more background information about this experiment, see the introduction in Experiment IX.

STRAINS

10-1   Ancestral strain—*MAT***a** prototroph
10-2   Evolved strain 1—derived from 10-1
10-3   Evolved strain 2—derived from 10-1
10-4   *MAT*α mating-type tester—XT1-20A—leu- ura- ade- *sst2 MAT*α
         (exact genotype not known)
10-5   *MAT***a** mating-type tester—RC634a—*ade2 his6 met1 ura1 can1 cyh1
         rme sst1-3 MAT***a**

EXPERIMENTAL PROCEDURE

Half of the class will receive a set of segregants from a backcross with strain 10-2, and the other half will receive segregants from the backcross with strain 10-3. We do not have as many markers segregating as you do in the tetrad analysis lab, and thus, we will only have the mating type to determine whether we have true tetrads. These particular strains are prototrophs, so we will use a mating-type test different from that used in previous experiments. Instead, we will pin the segregants onto a lawn of cells that cannot process the mating-type pheromones. The lawn cells will arrest the cell cycle if the patch of segregant cells is of the opposite mating type, creating a halo around the patch. We will sequence the best set of libraries from each cross.

▶ *Day 1*

Thaw a 96-well plate as follows: Remove the foil from the frozen plate. Replace the plastic cover and let it thaw slowly at room temperature, swabbing condensation with an ethanol-saturated KimWipe as necessary.

Fill a new round-bottom 96-well plate with 150 µL of YPD. Fill all wells with YPD even though not all wells of the frozen plate contain cells. These wells will serve as a negative control to detect contamination. Set up pinner and ethanol bath for flaming.

Flame pinner and allow to cool. Place in the left half of a thawed 96-well plate and move around to gently resuspend the cells, but without splashing. Place pinner in YPD-filled 96-well plate. Return pinner to same wells in source plate. Place pinner on YPD agar plate, making sure that all pins are evenly touching the surface.

Clean pinner with ethanol and flame again before moving on to another set of wells and repeating the process.

Place the new 96-well plate in the incubator on a flat surface (no shaking). Replace the original 96-well plate in −80°C freezer, after covering with foil tape and the plastic lid (see Techniques and Protocols 14).

▶ *Day 2*

Check that cell cultures are growing in your 96-well plate (and not in your negative control wells). TA will start 5 mL of overnight culture of halo tester strains.

▶ *Day 3*

Prepare halo lawns: Dilute the overnight cultures of halo tester strains 1:10 with sterile H$_2$O. Spread 200 µL of diluted culture evenly over a YPD plate using glass beads. Incubate for 30 min at 30°C. Prepare one plate for each tester strain per 48 wells of segregants (four plates total).

Set up pinner and ethanol bath for flaming. Pin as described on Day 1 from the liquid 96-well plate to the halo lawn plates. Make sure to clean pinner with ethanol and flame (and allow to cool) between each transfer because placing it on the halo plate will contaminate it with the mating-type tester strain. Allow the plates to dry before placing in incubator.

Set up two falcon tubes: clumpy and not clumpy. Score your liquid 96-well plates for evidence of settling, a proxy for clumping, using the template in Appendix C. You may need to resuspend the wells and then allow them to resettle to see the phenotype clearly. You may even need to look at each segregant under the microscope to score the clumping phenotype.

Make sure to notice the phenotype of each parental strain and the mated diploid. Is the trait dominant or recessive (note there is some ambiguity to this conclusion since some clumping-related traits differ with ploidy)?

After scoring, pool the cultures according to phenotype. Resuspend each segregant by pipeting up and down. Transfer the contents into the tube corresponding to either the clumpy or not clumpy pool.

Prepare DNA from the pools using the Hoffman–Winston prep (see Techniques and Protocols 4). Measure DNA concentration with the Qubit fluorometer using Broad Range reagents (see Techniques and Protocols 16).

If you have not done so already, during your incubation steps, look at the strains (particularly the evolved, ancestral, and mated diploids) under the microscope. Do they have a phenotype difference? Note any interesting features of the clumps, such as how the cells are stuck together, their shape, and their polarity. These observations could help you determine good candidate genes in the sequencing data.

Also check the agar plates for any other phenotypes, such as colony morphology differences and invasive growth (note that this step may be done over several days, as these phenotypes may increase as the patches overgrow). Make sure to take photographs of any interesting phenotypes. To determine whether any cells have invaded the agar, gently run water over the surface of the plate and wipe the colonies off with a gloved finger. Colonies that have invaded the agar will still have cells visibly present embedded in the plate, whereas those that have not invaded will be completely washed away. Place the washed plate back in the incubator to allow any invaded cells to regrow, so they will be more visible. If you do see any phenotypes, do they cosegregate with the clumping phenotype?

Prepare sequencing library using the Nextera kit and instructions below. We will give you each a set of unique primers to allow for multiplexed sequencing of multiple samples in the same sequencing lane.

Note: *For good results, maintain rigorous cleaning standards throughout this protocol!*

Any protocol involving a polymerase chain reaction (PCR) runs the risk of contamination by unwanted DNA sources being amplified in the final product. As we will be preparing fresh genomic DNA for sequencing, we want to avoid any contamination up front. ***Please wipe down your bench and all pipetmen with 10% bleach. Open a fresh box of tips for this protocol and a fresh bottle of dH$_2$O and use filter tips where possible.***

DNA quantification and quality

1. If you have not already done so, quantify your Hoffman–Winston prepped DNA samples using the Broad Range reagents on the Qubit.

2. Purify each DNA sample using the Zymo purification kit. Each column can only handle <5 μg of DNA: Do not overload!

3. Requantify your two purified DNA samples using the Broad Range buffers on the Qubit.

4. Nanodrop your DNA to determine if the quality is good. The 260/280 ratio should be between 1.8 and 2.0.

5. Dilute your DNA samples to 2.5 ng/μL and confirm the concentration once again, this time using the Qubit HS kit.

6. Stop here for today.

▶ Day 4

DNA tagmentation

1. Place aliquots of the tagmentation buffer and tagmentation enzyme on ice.

2. Thaw your genomic DNA samples.

3. Invert buffer and enzyme tubes three to five times to mix.

4. Mix together the following in this order for each of your two DNA samples in separate PCR tubes:

Genomic DNA (2.5 ng/μL)	20 μL
Tagmentation buffer (TAG)	25 μL
Tagmentation enzyme (TAG enzyme)	5 μL

5. Pipette up and down 10 times to mix.

6. Place tubes in a 55°C heat block or PCR machine for 5 min.

7. Place on ice.

Clean-up of tagmented DNA with Zymo columns

Note: *Follow this protocol, not the one given by the manufacturer.*

1. Mix 50 µL of the tagmentation reaction with 180 µL of Zymo DNA-binding buffer. Pipette up and down 10 times to mix and pipette onto a Zymo column already placed in its 2-mL collection tube.

2. Spin at maximum speed for 30 sec and discard flow-through.

3. Wash with 200 µL of wash buffer and centrifuge at maximum speed for 30 sec. Discard flow-through. Repeat for a total of two washes.

4. Centrifuge at maximum speed for 30 sec to remove residual liquid.

5. Place Zymo columns in clean 1.5-mL microcentrifuge tubes. Pipette 25 µL of the resuspension buffer (RSB) directly onto the filter of the Zymo column. Allow it to stand for 1 min. Centrifuge at maximum speed for 30 sec to elute cleaned DNA.

PCR amplification

1. To sequence multiple samples in the same flow cell, each DNA sample must have unique index sequences, which are contained in the primers. We have assigned primers such that each group's samples will be unique and compatible with other groups. You will receive two pairs of primers (index primers 1 and 2; one starts with a 7 and the other with a 5) corresponding to your two DNA samples and your aliquots of the Nextera PCR master mix (NPM) and the PCR primer cocktail (PPC). Place on ice.

2. Invert tubes three to five times to mix. Quick spin each tube.

3. Set up two PCR tubes for your two DNA samples as shown below.
   Note: *Change gloves between working with different primers to prevent cross-contamination.*

Index 1	5 µL
Index 2	5 µL
Nextera PCR master mix	15 µL
PCR primer cocktail	5 µL
Genomic DNA	20 µL

4. Pipette up and down three to five times to mix.

5. Quick spin each tube or flick to collect liquid in the bottom.

6. Run the following PCR (It is ok to run this PCR overnight):

| 72°C | 3 min | |
98°C	30 sec	
98°C	10 sec	
63°C	30 sec	5 cycles
72°C	3 min	
10°C	Forever	

PCR clean-up

1. Aliquot AMPure beads (at room temperature) and freshly prepared 80% ethanol.

2. Quick spin or flick your PCR tubes to collect the liquid at the bottom of the tube.

3. Vortex the AMPure beads for 30 sec to ensure that they are resuspended.

4. Add 30 μL of AMPure beads to each PCR; mix by pipeting up and down 10 times.

5. Incubate at room temperature for 5 min.

6. Place the tubes on the magnetic plate for at least 2 min or until the supernatant has cleared. Prepare two clean 1.5 microcentrifuge tubes for collecting the purified DNA.

7. With the tubes still on the magnetic plate, remove the supernatant without disturbing the beads and discard. Look at your pipette before discarding the supernatant to make sure you have not aspirated any beads; if you have, then repeat Steps 6 and 7.

8. With the tubes still on the magnetic plate, add 200 μL of 80% ethanol to each tube without disturbing the beads.

9. Wait 30 sec and remove the ethanol once again making sure to avoid aspirating any beads.

10. Repeat for a total of two washes.

11. Use a P20 or P10 tip to remove any residual ethanol.

12. Allow the pellet to dry for 15 min.

13. Remove tubes from the magnetic plate.

14. Add 32.5 µL of resuspension buffer to each tube and pipette up and down 10 times to resuspend the pellet.

15. Incubate at room temperature for 2 min.

16. Place the tubes back on the magnetic plate and incubate for 2 min.

17. Remove 30 µL of the supernatant using P20 or P10 and place into the appropriately labeled clean microcentrifuge tubes.

18. Quantify your PCR product using the Qubit HS kit.

Give your samples to the instructor. The four samples (clumpy/not clumpy for each of the two crosses) that look the best in terms of phenotype clarity and library quality will be submitted to the core facility for sequencing on a MiSeq instrument.

Mating-type testing

Check your halo plates each day to make sure that they do not overgrow. Cross-check the a and α plates to confirm mating type. If a segregant generates a halo, does that mean it is the same or a different mating type than the tester?

▶ Day 5

Pool libraries (the instructor will do this, but the instructions are included here for your reference).

1. Quantify each sample for pooling using the Qubit.

2. Calculate the molarity of each sample using the following conversion:

   1 ng/µL = 3 nM

3. Normalize each sample with Qiagen buffer EB (Tris-Cl, pH 8.5, 0.1% Tween) to 4 nM.

4. Mix 5 µL of each sample together.

5. Give to sequencing facility.

▶ Day 8

Get data from the core facility. We will map the reads and call mutations for you. Briefly, reads were aligned to the sacCer3 reference sequence using BWA (Li and Durbin 2009) and default parameters for paired-end reads. Mapped reads were converted into a SAM file. A file containing uniquely mapped reads was generated from the original SAM file. A final filtered mpileup file was generated using samtools (Li et al. 2009) with a -C50 filter, as recommended by BWA.

For SNP calling, a filtered VCF file was generated using vcftools, from the filtered mpileup file after removing duplicate reads. Additional filtering (using custom scripts, although the same thing can be accomplished in Excel) removed variants that were called in both the evolved and ancestral strain, and annotated the final variant call list with respect to genome location and coding changes (Pashkova et al. 2013; http://depts.washington.edu/sfields/software/annotate/).

If you work with a core facility, you can frequently have them perform these steps for you (or pay them to). Load the BAM file in the Integrative Genomics Viewer (http://www.broadinstitute.org/igv/). Look at the location of each mutation call and evaluate its quality. Compare between the pools and with other groups to determine the causative mutations for each trait. How many additional mutations are present?

MATERIALS

Note: Amounts provided are the requirements for each pair of students.

Day 1	96-well plates
	YPD plates
Day 3	Hoffman–Winston reagents
	DNA Clean and Concentrator-5 columns (D4004 Zymo Research)
	YPD plates
	Halo tester strain cultures
Day 4	Nextera index kit (Illumina FC-121-1011)
	Nextera sample preparation kit (Illumina FC-121-1030)
	Ampure reagents (Beckman Coulter A63880)
	DNA Clean and Concentrator-5 columns (D4004 Zymo Research)

REFERENCES

Li H, Durbin R. 2009. Fast and accurate short read alignment with Burrows–Wheeler transform. *Bioinformatics* **25**: 1754–1760.

Li H, Handsaker B, Wysoker A, Fennell T, Ruan J, Homer N, Marth G, Abecasis G, Durbin R, 1000 Genome Project Data Processing Subgroup. 2009. The sequence alignment/map (SAM) format and SAMtools. *Bioinformatics* **25**: 2078–2079.

Pashkova N, Gakhar L, Winistorfer SC, Sunshine AB, Rich M, Dunham MJ, Yu L, Piper RC. 2013. The yeast Alix homolog Bro1 functions as a ubiquitin receptor for protein sorting into multivesicular endosomes. *Dev Cell* **25**: 520–533.

Barcode Sequencing and Comparative Functional Genomics

Many of the yeast strain collections contain unique sequence tags known as barcodes. In the design of the yeast deletion collection, each cassette is flanked by two such 20-bp barcodes, an "uptag" and a "downtag." These barcodes are themselves each flanked by common primers. All uptags, for example, can be amplified using the same primers. This allows for pooled experiments to be performed, in contrast to experiments where each strain is individually grown and scored. Most typically, the entire set of deletion strains is mixed together and subjected to growth in an environment that allows for differential growth of the various mutants. Samples are taken over time as the strains compete and grow. The barcodes are amplified out of each time point, and the relative representation of each sequence is used as a readout of the survival of the corresponding strain.

Other strain collections and experimental systems have also adopted the convention of barcoding. For example, the CEN-MoBY and 2μ-MoBY libraries contain all genes on barcoded CEN and 2μ plasmids, respectively (Ho et al. 2009; Magtanong et al. 2011). The "barcoder" yeast collection contains thousands of strains with unique sequences integrated at the *HO* locus, which allows a user to create their own barcoded libraries (Yan et al. 2008; Douglas et al. 2012). Next-generation sequencing libraries also utilize barcode sequences for subassembly methods and other applications.

Barcode sequences can be measured in several ways. Initially, Affymetrix microarrays containing each 20-base barcode were used. Additional microarray platforms have also been successful. As these studies have been carried out, various problems with the initial barcode sequences have been discovered; for example, some of the sequences are simply incorrect or are difficult to amplify. Next-generation sequencing has emerged as a useful approach to circumvent some of the problems with arrays, in that the exact barcode sequence is recovered as opposed to relying on the array design for a match (however, the genotype still needs to be linked to the barcode). The sequencing approach is called "bar-seq" (Smith et al. 2009) and relies on the same basic premise as using read depth to measure DNA copy number: The number of times the barcode is sequenced is proportional to the copy number. In turn, the barcode copy number is proportional to the strain frequency.

Because the barcode sequences are small, the polymerase chain reaction (PCR) amplifications should be performed with a little more care than usual. A solution with substantial mass of DNA will have very high molarity. This means that even cursory contamination of pipettes, gel boxes, etc., can be exponentially amplified and cause false signals. It is recommended that you isolate barcode PCR setup from post-PCR analysis to avoid contamination. In addition, clean all tools well and work in a low-DNA environment such as a UV-irradiated hood. Be particularly careful when making stocks of primers.

In this experiment, we will compete a new collection made in the strain background Σ1278b (3.2 polymorphisms/kb vs. the S288c reference genome). This strain background, unlike those derived from S288c, has the ability to undergo invasive growth, make pseudohyphae, and flocculate. These features have made it a popular strain background for studying those phenomena. The Boone lab has created a deletion collection in this background (Dowell et al. 2010; Ryan et al. 2012). We will compare the osmotic stress growth profiles of mutants in the BY and Σ1278b backgrounds in order to determine to what extent these phenotypes are conserved in the face of genetic variation.

STRAINS

11-1   BY deletion collection. *MATa ura3Δ0 leu2Δ0 his3Δ1 lyp1Δ can1Δ:: LEU2-MFA1pr-HIS3 xxx::KanMX*

11-2   Σ1278b deletion collection. *MATα ura3Δ leu2Δ his3::hisG can1Δ:: STE2pr-Sp_his5 lyp1Δ::STE3pr-LEU2 xxx::KanMX*

EXPERIMENTAL PROCEDURE

One half of the class will work with the BY collection and the other half with the Σ1278b collection. We will allow each collection of strains to grow competitively for 3 d, diluting the cultures back once a day. This should allow the more fit strains to increase in relative frequency and the sicker strains to decrease measurably. One aliquot of each strain collection will be subjected to osmotic stress while a control sample will grow without osmotic stress. We will then sequence the barcodes from the starting point and three time points for each competition (seven samples total/two-person team). Because the MiSeq cannot handle all the samples, the best libraries for each pool + condition will be sequenced and everyone will analyze the resulting data set. You will then select one mutant for validation by growth curve.

▶ Day 1

Try to complete these steps early in the day to leave time for competitive growth before collecting samples on the following day

Thaw an aliquot of your assigned library (BY or Σ1278b) at room temperature and resuspend the cells by pipetting up and down.

Pipette one-third of the tube into 50 mL of YPD and another one-third into 50 mL of YPD + 1.5 M sorbitol.

Place flasks in the 30°C shaker.

Harvest the remaining one-third of the starting pool by spinning down and freezing the pellet. This will be your time-0 sample. Record the time today and over the following days in the table provided.

Barseq time course data	Date	Time	OD	Doublings elapsed
Time 0				
~24 h YPD				
~24 h YPD + sorbitol				
~48 h YPD				
~48 h YPD + sorbitol				
~72 h YPD				
~72 h YPD + sorbitol				

▶ Day 2

Try to complete these steps ~24 h after time 0.

Measure the OD of the culture. If you are using Σ1278b, work quickly before the cells settle. Dilute back into a new flask of YPD or YPD + sorbitol at an OD of 0.05. Put the new flasks back in the shaker. Compare the growth of your sample to that of another group working with the other strain background.

Harvest the remaining culture by centrifuging and freezing the pellet.

Use the OD measurements to calculate the number of doublings the population has gone through. (Note that if you are using Σ1278b, work quickly before it settles.) Record in the data table.

▶ Day 3

Repeat the procedure from Day 2. The culture should be more dense today, so harvest only 10 mL. You can save the rest of the culture as well as a backup if desired.

▶ Day 4

Measure the OD of the culture. You do not need to dilute back today. Just harvest as on Day 3.

Prepare DNA from the seven frozen samples. Use the Hoffman–Winston prep starting from the pellet step (see Techniques and Protocols 4). Check DNA concentration using the Qubit fluorometer (see Techniques and Protocols 16). Clean up DNA with a purification column. Measure concentration with Qubit.

We will distribute primers unique for each group and each sample. Because the barcode libraries are much lower complexity than the whole-genome sequences from Experiment X, we will be able to sequence more samples in a single MiSeq lane. However, we still cannot fit everybody, so we will just select the best samples and distribute the data to everyone to analyze.

Each time point for YPD and YPD + 1.5 M sorbitol will require a unique set of primers, resulting in seven primer pairs in total. The reverse primer is the same for all the PCRs and contains homology with the KanMX cassette. The forward primer contains homology with the common uptag priming site, as well as a unique index sequence that will be used to determine the sample and time point. Therefore, it is important to record the primer number for each sample. It is also important to run a no-DNA control for each set of primers to detect contamination. You will set up a total of 14 PCRs.

Bar-seq PCR

Sample number	Sample condition	Time	Forward primer	Reverse primer
1	Starting pool	Time 0		
2	YPD	~24 h		
3	YPD + sorbitol	~24 h		
4	YPD	~48 h		
5	YPD + sorbitol	~48 h		
6	YPD	~72 h		
7	YPD + sorbitol	~72 h		

PCRs

	Sample	No DNA control
Fast start buffer	10	10
dNTP (10 mM each)	1.2	1.2
Fast-start *taq*	0.8	0.8
Oligo1 (10 µM)	8	8
Oligo2 (10 µM)	8	8
DNA (~100 ng/µL)	1	0
H$_2$O to 100 µL		72

PCR program

94°C	3 min	
94°C	30 sec	
55°C	30 sec	25 cycles
72°C	30 sec	
72°C	3 min	
4°C	15 min	

Clean PCR product with MinElute PCR purification kit. Elute with 10 μL of elution buffer (EB).

▶ Day 5

Save the no DNA controls to run on gel at a later date. Proceed with the seven PCRs done with the DNA template added.

Purify the PCR product using the Ampure kit.

Acclimate the Ampure beads to room temperature for 5–10 min.

Shake the beads.

Add 1.8 μL of beads per μL of PCR. Pipette up and down 10 times to mix. Incubate at room temperature for 5 min.

Place on magnet plate for 2 min. Beads will accumulate on one side of the tube.

Remove supernatant. Be careful not to remove beads because the double-stranded DNA is bound to beads.

Wash with 200 μL of 70% ethanol. Wait for 30 sec.

Remove 70% ethanol. Be careful not to disturb the beads.

Repeat the ethanol wash.

Dry beads for 15 min.

Remove PCR tube from the magnet plate.

Elute with 30 μL of buffer EB. Resuspend by pipetting up and down 20 times. Incubate at room temperature for 2 min.

Place PCR tube back onto the magnet for 2 min.

Remove supernatant from PCR tube and place into clean 1.5-mL tube. (Be careful not to grab any beads.)

Measure DNA yield using the Qubit Broad Range reagents (see Techniques and Protocols 16).

The instructor will collect the samples and check quality control by running them on a gel. Samples that contain PCR products with an average length of 375 bp successfully amplified the barcodes and will be sequenced. We will pool equal amounts of DNA from each sample for sequencing.

▶ Day 9

Get the sequencing data from the core facility.

We will do the initial analysis of the sequencing data. This involves splitting the file by index reads to divide the individual samples out and mapping the sequences against the master barcode file to match them back to the strain from which they were amplified. The scripts for this analysis are available in the supplement to Payen et al. (2015).

The rest of the analysis is up to you.

Exclude barcodes with very low read numbers. You may want to keep track of these genes in a separate file.

Normalize each sample by dividing the number of reads for each barcode by the total reads in that sample.

Normalize by the initial pool abundance by taking the log ratio between each time point and the starting pool.

Rank the data by ratio and look at mutants that grew particularly well or poorly.

Compare the rankings between the samples. How many deletion strains change in relative abundance? Do any disappear entirely over the time course? Are there deletions that change their behavior in sorbitol? How about between the two strain backgrounds?

Use the SGD tools you learned about to try to discern patterns about these sets of genes.

▶ Day 13

It is generally a good idea to validate interesting findings from genomic-scale experiments. Pick one mutant you would like to retest by growth curve and email the strain name to the TA. We will pull the strains from the deletion collection for you to validate.

You can pick whatever strain you want, based on whatever analysis you like. You can pick a candidate gene based on the literature, or you can choose a strain on the basis of your data. For example, if you find a mutant that is sensitive to osmotic stress in one strain background but not the other, it could be a good strain to validate. Alternatively, you could test a hypothesis that a gene known to regulate osmotic stress response behaves similarly in both strain backgrounds. We will give you both the Σ1278b and BY knockouts, if available.

▶ Day 15

The deletion collections are great resources, but they also can be tricky to work with. One common problem is cross-contamination between wells. This is an especially pernicious problem for slow-growing strains, which may be quickly outcompeted by their better-growing neighbors even starting from only scant contamination. To ensure that you are testing the correct strain, you can sequence the barcode, just as we did for the competition experiment. Instead of using high-throughput sequencing for this, we will use Sanger sequencing of PCR products.

For simple PCRs like this, you can frequently get away with performing PCR directly from the colony, skipping the need to make DNA. Just as a warning: Colony PCR does not always work, particularly for certain finicky PCRs.

COLONY PCR

On the bottom of each plate, circle a well-isolated colony.

Perform colony PCR according to Techniques and Protocols 17.

After you have set up your PCR, inoculate an overnight culture using the exact same colony you chose for PCR. This ensures that you are sequencing the barcode for the same isolate whose growth curve is being performed.

Purify the PCR product with the Zymo column cleanup (follow the kit's instructions) and elute with 25 µL of H_2O.

Measure the DNA concentration of 1.5 µL using the Nanodrop. Give the DNA to the instructor to submit for Sanger sequencing.

▶ Day 16

Begin this experiment early in the day. You will calculate the growth rate of each strain using a standard growth curve.

Dilute each overnight culture back to an OD of 0.05 in a flask of 50 mL of media, with one flask containing YPD and the other YPD + sorbitol. Remove a small sample and check the OD at approximately hourly intervals throughout the day. Initially, you may need to remove 1 mL to get an accurate OD reading, but as the culture gets denser, you may remove less and dilute it for the OD measurement. Record your measurements in the table below. Keep making measurements until you leave the lab tonight. Allow the culture to continue growing overnight.

Time and date	OD wt BY in YPD	OD Δ BY in YPD + sorbitol	OD wt Σ1278b in YPD	OD Δ Σ1278b in YPD + sorbitol

▶ Day 17

Take a final time point for your growth curve. Use a graph to record the data. During maximal growth, density will increase exponentially (only this phase of growth is graphed below).

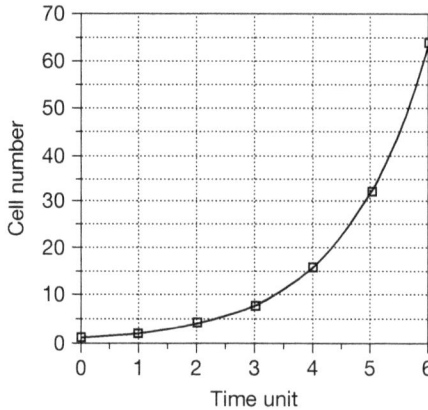

You can see this better by graphing the data on a semilog scale (linear time but log cell density).

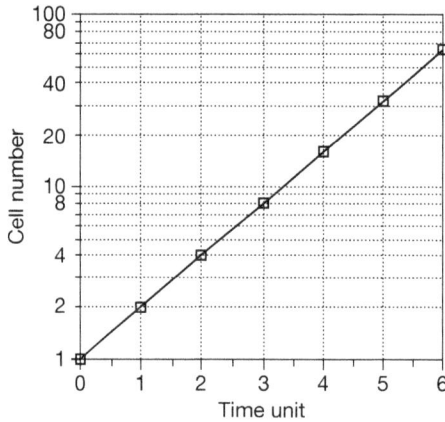

Locate the linear section of the growth curve on the semilog plot. Although stress resistance/sensitivity effects can also frequently be seen in lag phase, saturation density, and stationary phase viability, most of the time when you measure growth rate, the maximal growth rate is what is being referred to. Calculate the doubling time of the culture either by regressing a line to just this segment of the curve or by simply determining the time it takes for the culture to double in density by inspection. Also examine the growth curves for other differences in lag length and saturation density. Compare the control to the deletion strain for each strain background, and compare between strain backgrounds. Have you reproduced the result from the pooled competition experiment?

You will also receive your Sanger sequencing data back today. We will give you a reference sequence table to use to locate the barcode sequence within your data. Are the strains what you thought they were?

MATERIALS

Note: Amounts provided are the requirements for each pair of students.

Day 1   1 Aliquot BY or Σ1278b library
                 50 mL of YPD
                 50 mL of YPD + 1.5 M sorbitol

Day 2   50 mL of YPD
                 50 mL of YPD + 1.5 M sorbitol

Day 3   50 mL of YPD
                 50 mL of YPD + 1.5 M sorbitol

Day 4   DNA prep and PCR reagents
                 Uniquely indexed Barseq-F primer:
                 CAAGCAGAAGA CGGCATACGAGAT[6-base-index]
                    GCGCTCCGAGCGGATGTCCACGAGGTCTCT
                 Common Barseq-R primer:
                 AATGATACGGC GACCACCGAGA TCTACACGGCC
                    GTCGACAATTCAACGCGTCTGTGAGGGGAGCG
                 MinElute columns (Qiagen 28006)

Day 5   Ampure reagents (Beckman Coulter A63880)
                 Custom Illumina sequencing primers:
                    Sequencing primer for read 1: CCGGGGATCCGTCGACCTGCA
                       GCGTACG
                    Index read primer: AGAGACCTCGTGGACATCCGCTCGGAGCGC

Day 15   Colony PCR reagents
                 Barcode Sanger confirmation primers:

Forward Oligo1: GATGTCCACGAGGTCTCT
Reverse Oligo2: ACAATTCAACGCGTCTGTGA

Day 16   2 × 50 mL of YPD
             2 × 50 mL of YPD + 1.5 M sorbitol

REFERENCES

Douglas AC, Smith AM, Sharifpoor S, Yan Z, Durbic T, Heisler LE, Lee AY, Ryan O, Göttert H, Surendra A, et al. 2012. Functional analysis with a barcoder yeast gene overexpression system. *G3 (Bethesda)* **2:** 1279–1289.

Dowell RD, Ryan O, Jansen A, Cheung D, Agarwala S, Danford T, Bernstein DA, Rolfe PA, Heisler LE, Chin B, et al. 2010. Genotype to phenotype: A complex problem. *Science* **328:** 469–469.

Ho CH, Magtanong L, Barker SL, Gresham D, Nishimura S, Natarajan P, Koh JL, Porter J, Gray CA, Andersen RJ, et al. 2009. A molecular barcoded yeast ORF library enables mode-of-action analysis of bioactive compounds. *Nat Biotechnol* **27:** 369–377.

Magtanong L, Ho CH, Barker SL, Jiao W, Baryshnikova A, Bahr S, Smith AM, Heisler LE, Choy JS, Kuzmin E, et al. 2011. Dosage suppression genetic interaction networks enhance functional wiring diagrams of the cell. *Nat Biotechnol* **29:** 505–511.

Payen C, Sunshine AB, Ong GT, Pogachar JL, Zhao W, Dunham MJ. 2015. Empirical determinants of adaptive mutations in yeast experimental evolution. biorXiv doi: http://dx.doi.org/10.1101/014068

Ryan O, Shapiro RS, Kurat CF, Mayhew D, Baryshnikova A, Chin B, Lin ZY, Cox MJ, Vizeacoumar F, Cheung D, et al. 2012. Global gene deletion analysis exploring yeast filamentous growth. *Science* **337:** 1353–1356.

Smith AM, Heisler LE, Mellor J, Kaper F, Thompson MJ, Chee M, Roth FP, Giaever G, Nislow C 2009. Quantitative phenotyping via deep barcode sequencing. *Genome Res* **19:** 1836–1842.

Yan Z, Costanzo M, Heisler LE, Paw J, Kaper F, Andrews BJ, Boone C, Giaever G, Nislow C. 2008. Yeast Barcoders: A chemogenomic application of a universal donor-strain collection carrying bar-code identifiers. *Nat Methods* **5:** 719–725.

Genomic Modifications with PCR Products

The genome of *Saccharomyces cerevisiae* can be modified easily due to its propensity to repair double-stranded DNA breaks by homologous recombination. Transformation of yeast with appropriate DNA constructs leads to targeted deletion, substitution, and modification of genes. Polymerase chain reaction (PCR) permits the generation of constructs for these alterations in a single step, thereby skipping traditional, time-consuming cloning in bacteria. Thus, genome modifications with PCR products are referred to as one-step gene replacements or one-step gene modifications. This section supplements the manual by describing PCR protocols for one-step gene replacement in (1) de novo gene disruption, (2) gene disruption using the yeast deletion collection, and (3) generating protein fusions. For a comprehensive review of PCR-based approaches, see Maeder et al. (2007) and Gardner and Jaspersen (2014).

Approximately 40–60 bp of homology on each end of a PCR product are required for targeting homologous recombination with the genome. These can be acquired either with templates that already contain genomic homology or by using PCR primers with 40–60-bp tails that append the homology. A large collection of PCR templates with a variety of marker genes and useful protein modules (e.g., green fluorescent protein [GFP]) are now available from sources such as the European Saccharomyces Cerevisiae Archive for Functional Analysis (EUROSCARF; http://web.uni-frankfurt.de/fb15/micro/euroscarf) and Addgene.com (http://www .addgene.org/). Many of the templates contain universal binding sites for PCR primers such that a single set of oligonucleotides can be used to amplify different modules or marker genes for modification of a given gene (for examples, see Wach 1996; Longtine et al. 1998; Goldstein and McCusker 1999; Knop et al. 1999; Janke et al. 2004; Sheff and Thorn 2004).

Even with 40–60 bp of flanking homology, mistargeting occurs with measurable frequency, particularly if the integration cassette carries sequences homologous to the yeast genome. For example, attempts to replace genes in strain W303 with *TRP1* will frequently yield gene conversion of the strain's *trp1-1* allele. Marker genes that have limited homology with the *S. cerevisiae* genome have been developed to avoid such off-target events. The *his5*⁺ gene of *Schizosaccharomyces pombe* and the *URA3* gene of *Kluyveromyces lactis* are transcribed in *S. cerevisiae* where they complement the *HIS3* and *URA3* genes despite significant divergence in sequence. Dominant antibiotic resistance genes that bear no homology with the budding yeast genome have also been developed. The bacterial *kanʳ*, *hphʳ*, and *natʳ* genes confer resistance in yeast to G418, hygromycin B, and ClonNat, respectively. In each case, the open reading frame (ORF) of the resistance gene was flanked by the *TEF1* promoter and terminator from a related yeast, *Ashbya gossypii*, to yield an "MX" allele (*KANMX*, *HYGMX*, *NATMX*) (Wach et al. 1994). This makes it easy to swap MX markers with one another by homologous recombination, as described below. Many cassettes are also flanked by *loxP* sites to permit subsequent removal by expression of the Cre recombinase (Güldener et al. 1996).

It is good practice to transform diploid yeast strains during one-step gene replacement procedures. This reduces selective pressure for suppressors in the event that the resulting gene modification is required for viability or robust growth. Upon sporulation, haploid strains of both mating types are recovered, which can be useful for subsequent strain manipulations. Furthermore, unexpected spontaneous phenotypes that arise during the transformation procedure can be identified and avoided during tetrad dissection. Because of the short time frame of the course, however, we transform haploid strains directly.

Proper targeting of PCR fragments can sometimes be screened by a diagnostic property of the transformed cells. Gain of a marker gene phenotype is not sufficient to rule out off-target events. A more rigorous approach utilizes PCR. Primers should be designed to confirm both the gain of new genomic sequences and the loss of the original sequences. For example, one primer that binds adjacent to the modified locus and one that binds within the added sequence (e.g., the *AgTEF1* sequences) will detect additions. Conversely, a primer that binds adjacent to the modified locus and one that binds within the deleted sequence will detect the absence of modifications.

A. De Novo Gene Disruption by One-Step Gene Replacement

Generation of PCR Products

1. Design PCR primers that amplify the marker gene to be used and that append a 40–60-nucleotide homologous sequence upstream and downstream from the ORF to be deleted (Fig. 1). When designing full gene deletions, start and stop codons are frequently left intact. In some cases, only a partial gene deletion is desired. When purchasing custom oligonucleotides, it is not necessary to have them purified by high-performance liquid chromatography (HPLC)

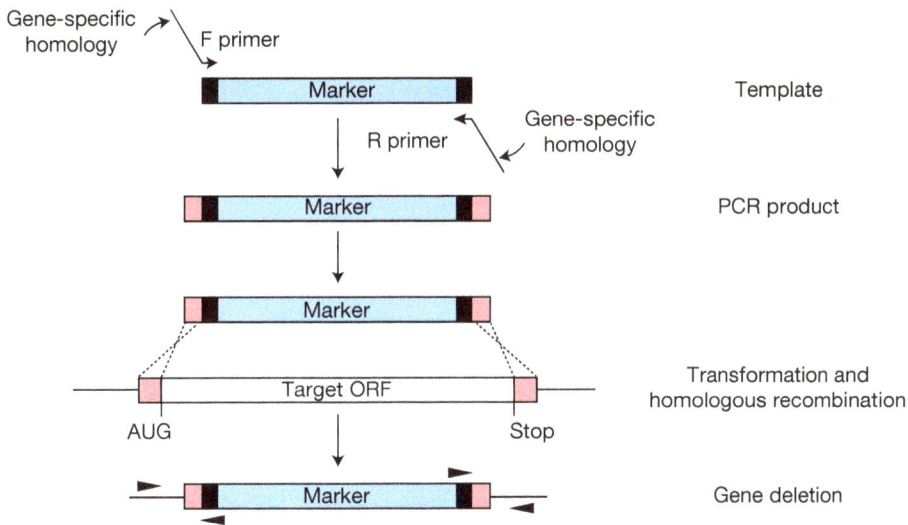

FIGURE 1. De novo gene disruption by one-step gene replacement. The F and R primers bind the template of the marker gene and contain tails of 40–60 bp that have homology with the target. F, ORF and R binding sites are shown in black and added homology in the PCR product is shown in red. Correct integration at the target ORF can be confirmed by PCR with the primers shown as arrowheads in the last step of the diagram.

or polyacrylamide gel electrophoresis (PAGE). In the example described in Figure 1, the single set of primers below could be used with the gene replacement templates described in Longtine et al. (1998), Goldstein and McCusker (1999), and others.

Forward primer:
5′-40 bp upstream of ORF-ATG-cggatccccgggttaattaa-3′
Reverse primer:
5′-40 bp downstream from ORF-stop codon-tcgatgaattcgagctcgt-3′

2. Go to Section D for PCR protocol.

B. Gene Disruption by One-Step Gene Replacement Using the Yeast Deletion Collection

In 1996, a consortium of labs systematically deleted all 6000 yeast ORFs to generate a complete collection of yeast knockouts using the *KANMX* marker. As part of the deletion process, each gene deletion was uniquely "tagged" with one or two 20-mer sequences that serve as a "barcode" to recognize the deletion by microarray or whole-genome sequencing. These deletions are available as heterozygous diploids, as well as haploids and homozygous diploids, at least diploids for the nonessential genes. In addition to being useful for experiments, the deletion collection provides a convenient set of PCR templates for creating gene deletions in new strains. Importantly, these premade deletions permit the generation of PCR products that carry long stretches of flanking homology. Commonly, PCR primers are designed to bind about 200 bp upstream and downstream from an existing *KANMX*-marked deletion. The efficiency of targeting by such a PCR product is vastly enhanced by the long flanking spans of homology.

Generation of PCR Products

1. Design PCR primers, each with a melting temperature of 56°C, that bind ~200 bp upstream and downstream from the gene deletion of interest (Fig. 2).

2. It is often desirable to swap one MX marker for another in a modified strain (replacing *KANMX* with *NATMX*, for example). Because of the generic structure of MX markers, this can easily be achieved by gene replacement with a PCR product bearing just the new marker gene flanked by the *AgTEF1* promoter and terminator. Universal primers for marker swapping are provided below. When swapping markers, confirm loss of the original marker by replica plating.

   a. AGTEF1p primer   5′-ccttgacagtcttgacgtgc-3′

   b. AGTEF1t primer   5′-cgcacttaacttcgcatctg-3′

3. Go to Section D for PCR protocol.

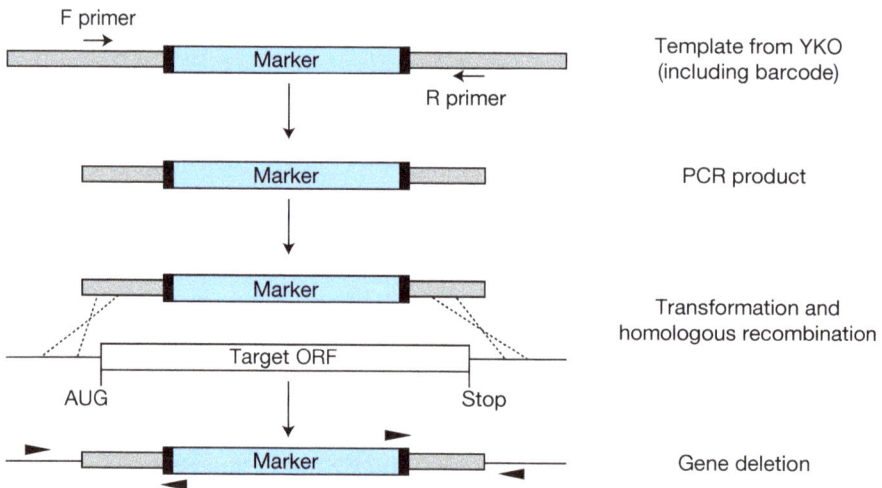

FIGURE 2. Gene disruption by one-step gene replacement using the yeast deletion collection. Primers for PCR amplification of genomic DNA of an existing deletion, such as a knockout from the yeast deletion collection, are designed to anneal ~200 bp upstream of the ATG and downstream from the stop codon. Correct integration at the target ORF can be confirmed by PCR with the primers shown as arrowheads in the last step of the diagram.

C. Generating Protein Fusions by One-Step Gene Modification

Tagging proteins with short peptides or small functional polypeptides facilitates analysis of proteins, particularly in yeast where making protein fusions is quick and simple. A common modification involves the addition of antigenic peptides, often as tandem repeats, for which antibodies are commercially available. These include the highly antigenic HA peptide from the hemagglutinin protein of influenza virus, a MYC peptide from the c-Myc protein, and the FLAG epitope, which was designed specifically for protein tagging. A second class of modifications involves fusing polypeptides with function. In addition to serving as epitopes for matching antibodies, these polypeptides confer great utility both inside and outside the cell. For example, autofluorescent polypeptides, such as GFP and mCherry (a nonoligomeric variant of red fluorescent protein), are used to localize fusion proteins in vivo. Other fused polypeptides provide handles for affinity chromatography of protein complexes. Examples include glutathione-S-transferase (GST) and tandem affinity purification (TAP). Collections with nearly every ORF carboxy-terminally tagged with GFP and TAP are available from commercial sources.

The most common way to make protein fusions is by one-step gene modification with PCR fragments. Numerous cassettes with different tags adjacent to selectable marker genes are available from EUROSCARF and Addgene (see Longtine et al. 1998; Knop et al. 1999; Janke et al. 2004; Sheff and Thorn 2004). The modules can be amplified by PCR using oligonucleotides that contain 40–60-bp extensions with sequence-specific information to direct integration to the gene of interest. Fusions can be made to both ends of the gene (and even within ORFs), but appending sequences to the 3′ end is simplest and most common. Modifications at the 5′ end of a gene typically require subsequent removal of the marker to rejoin the endogenous promoter with the modified ORF. In these cases, subsequent application of Cre recombinase can remove marker genes that are flanked by *loxP* sites, as described above.

Accurate integrations should be confirmed by PCR and/or by functional assays for the added peptide, such as microscopy for fluorescent proteins or immunoblotting for the epitopes. All protein fusions should be tested for function of the gene of interest. This is best accomplished by demonstrating complementation of *all* mutant phenotypes associated with the deletion of the gene. Growth on rich media at 30°C does not necessarily mean that a chimeric gene is fully functional.

Generation of PCR Products

1. Design PCR primers with 40–60 bp of sequence upstream/downstream from the target site of integration, followed by specific sequences that recognize

the tagging module (Fig. 3). In the example provided, the single set of primers below could be used with templates for carboxy-terminal tagging with fluorescent proteins described in Sheff and Thorn (2004).

> Forward primer:
> 5′-40 bp upstream of stop codon of ORF-ggtgacggtgctggttta-3′
> Reverse primer:
> 5′-40 bp downstream from stop codon of ORF-cgatgaattcgagctcg-3′

2. Go to Section D for PCR protocol.

D. PCR Protocol for Gene Modifications

1. Purchase primers and resuspend in H_2O at a final concentration of 25 μM.

2. Choose a thermostable polymerase suitable for the protocol. *Taq* does not possess an editing activity and therefore is prone to a higher rate of misincorporation. Nevertheless, the enzyme is cheap, robust, and widely available. *Taq* is suitable for making gene disruptions. When making fusion genes, a thermostable proofreading DNA polymerase would be more appropriate (e.g., Phusion or Vent from New England Biolabs). To amplify long DNA targets (>5 kb), it is common to use a mixture of *Taq* with a proofreading

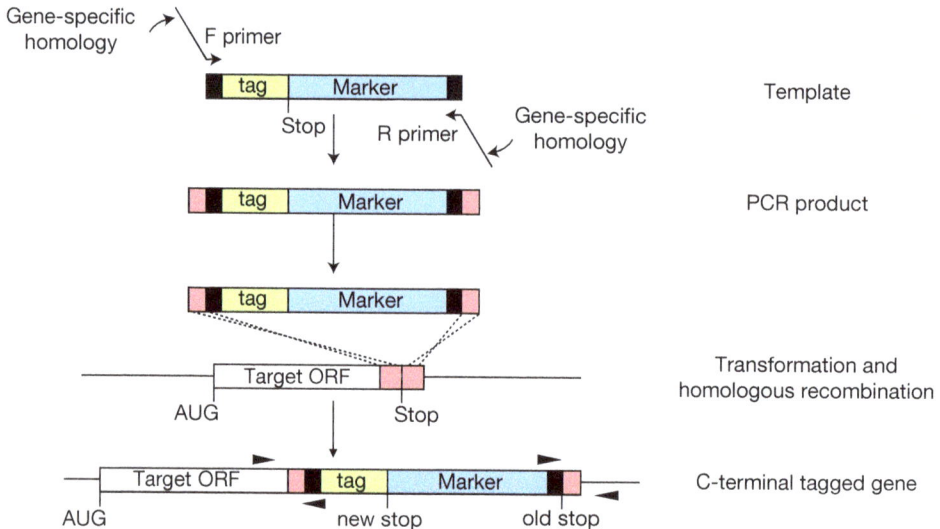

FIGURE 3. Generating protein fusions by one-step gene modification. Primers for PCR amplification of the cassette are designed to contain 40–60 bp of sequence upstream and downstream from the stop codon. It is important that the F primer maintains the ORF and omits the stop codon of the targeted gene. Correct integration can be verified by PCR with the primers shown as arrowheads in last step.

polymerase (e.g., OneTaq, New England Biolabs). Importantly, use each polymerase at a temperature optimal for the enzyme, as specified by the manufacturer.

3. Program the thermocycler. Annealing temperature will vary depending on the melting temperature of the sequence of the primer that binds the template in the first round. Extension time will vary depending on the length of the amplified sequence and the properties of the DNA polymerase, as described by the manufacturer. The following parameters were selected based on using Q5 DNA polymerase (NEB) and an extension length of 2 kb (allow 20–30 sec extension per kb when using Q5). These conditions are appropriate for the PCRs described in Experiments I and II.

Step	Temperature (°C)	Time (sec)	Number of cycles
Denaturation	98	30	1
Denaturation	98	10	
Annealing	56	30	30
Extension	72	60	
Final extension	72	120	1
Hold	4	Infinite	1

4. Set up the following PCR on ice in a 0.2-mL PCR tube by adding the following reagents in the following order. Do not add polymerase until immediately before transferring the samples to the thermocycler.

DNA template (~10 ng)	1 µL
25 µM Forward primer	1 µL
25 µM Reverse primer	1 µL
5× Q5 Polymerase buffer	10 µL
10 mM dNTPs (NEBL N0447S)	1 µL
H_2O	35.5 µL

Mix by pipetting.

Q5 DNA polymerase (NEBL M0491S)	0.5 µL

5. Confirm PCR. Pour a 1% agarose-TBE gel by mixing 0.5 g of agarose with 50 mL of 1× TBE in a 250-mL bottle or Erlenmeyer flask. Microwave until boiling (~1 min). *Caution: Boiling agarose erupts easily, spilling out of the container.* Mix gently until all agarose dissolves. Allow solution to cool until bottle can be held comfortably by hand. Pour solution into the gel form and add comb. Allow

it to solidify for 15–30 min until agarose becomes slightly opaque. Remove comb and place gel in gel running box. Submerge in 1× TBE.

6. Add 2 µL of the PCR to 10 µL of 1× loading buffer. Load sample into one well of an agarose gel. In an adjacent well, load 10 µL of diluted DNA ladder (see below). Run at ~80 V for 30–60 min.

7. Stain gel by gently rocking in H_2O plus SYBR Safe dye (10-µL stock/100 mL).

8. Visualize DNA using UV box and photograph to record for lab book.

9. If the PCR yields a robust band on the gel, use 25% of the product for each yeast transformation.

REAGENTS

10× TBE

108 g of Tris-base
55 g of boric acid
40 mL of 500 mM EDTA (pH 8)
H_2O to 1 L

Diluted DNA Ladder

5 µL of stock ladder (NEB 3232)
40 µL of 6× loading buffer
155 µL of H_2O

6x Loading Buffer

30% glycerol in H_2O
1% bromophenol blue
1% xylene cyanol

REFERENCES

Gardner JM, Jaspersen SL. 2014. Manipulating the yeast genome: Deletion, mutation, and tagging by PCR. *Methods Mol Biol* **1205:** 45–78.

Goldstein AL, McCusker JH. 1999. Three dominant drug resistance cassettes for gene disruption in *Saccharomyces cerevisiae*. *Yeast* **15:** 1541–1553.

Güldener U, Heck S, Fielder T, Beinhauer J, Hegemann JH. 1996. New efficient gene disruption cassette for repeated use in budding yeast. *Nucleic Acids Res* **24:** 2519–2524.

Janke C, Magiera MM, Rathfelder N, et al. 2004. A versatile toolbox for PCR-based tagging of yeast genes: New fluorescent proteins, more markers and promoter substitution cassettes. *Yeast* **21**: 947–962.

Knop M, Siegers K, Pereira G, Zachariae W, Winsor B, Nasmyth K, Schiebel E. 1999. Epitope tagging of yeast genes using a PCR-based strategy: More tags and improved practical routines. *Yeast* **15**: 963–972.

Longtine MS, McKenzie III A, Demarini DJ, Shah NG, Wach A, Brachat A, Philippsen P, Pringle JR. 1998. Additional modules for versatile and economical PCR-based gene deletion and modification in *Saccharomyces cerevisiae*. *Yeast* **14**: 953–961.

Maeder CI, Maier P, Knop M. 2007. A guided tour to PCR-based genomic manipulations of *S. cerevisiae* (PCR-targeting). In *Methods in microbiology* (ed. Stansfield I, Stark MJR), vol. 36, pp. 55–78. Academic Press, New York.

Sheff MA, Thorn KS. 2004. Optimized cassettes for fluorescent protein tagging in *Saccharomyces cerevisiae*. *Yeast* **21**: 661–670.

Wach A. 1996. PCR-synthesis of marker cassettes with long flanking homology regions for gene disruptions in *S. cerevisiae*. *Yeast* **12**: 259–265.

Wach A, Brachat A, Pohlmann R, Philippsen P. 1994. New heterologous modules for classical or PCR-based gene disruptions in *Saccharomyces cerevisiae*. *Yeast* **12**: 259–265.

High-Efficiency Yeast Transformation

PROCEDURE

1. Start an overnight culture in YPD or selective media and shake overnight at an appropriate temperature for the strain. The goal is to obtain 50 mL of mid-log culture the following day. To this end, inoculate a 5-mL culture heavily and plan to back-dilute in the morning or inoculate a 50-mL culture lightly and hope for the best.

2. Measure OD_{600} in the morning.

   a. If OD_{600} is >1.0, dilute cells back to 0.1 and grow for 4–6 h.

   b. If OD_{600} is 0.2–1.0, use immediately or dilute for use later in the day.

   c. If OD_{600} is less than 0.2, continue growing cells.

   Note: Because of light scattering, accurate OD measurements can only be made between the range of 0.1 and 0.3 (note that linear range may vary from one spectrophotometer to the next). Dilute accordingly before making measurements.

3. Centrifuge the culture in a 50-mL conical tube for 3–5 min at 3000g (2500 rpm) in a tabletop centrifuge.

4. Thaw single-stranded carrier DNA. Boil for 3 min and cool on ice.

   Note: It is not necessary or desirable to boil the carrier DNA every time. Keep a small aliquot in a −20°C freezer and boil after three or four freeze/thaw cycles.

5. Pour off media, pipette off residual liquid, and resuspend cell pellet in 10 mL of 1× LiOAc buffer by vortexing.

6. Centrifuge the cells for 3 min at 3000g (2500 rpm).

7. Pour off LiOAc buffer, pipette off residual liquid, and resuspend pellet in 0.5–1 mL of 1× LiOAc buffer.

8. In a 1.5-mL Eppendorf tube, mix

   a. 1–5 µg of plasmid DNA or 25% of an efficient PCR in no more than 10 µL of total volume. Concentrate DNA by ethanol precipitation if necessary.

   b. 10 µL of 10 mg/mL carrier DNA.

   c. 100 µL of cells in 1× LiOAc buffer.

   Note: Include a no-DNA control.

9. Add 280 µL of PEG solution to each tube. Vortex or pipette vigorously to mix the PEG with other components. Incubate 20–45 min at room temperature.

10. Add 43 µL of DMSO to each tube. Vortex or pipette to mix thoroughly.

11. Incubate 5–15 min at 42°C. Chill on ice immediately afterward.

12. Centrifuge for 2 min at half speed in a microcentrifuge. Aspirate off liquid.

13. Optional washing step to remove residual transformation buffer: Add 0.5 mL of YPD or sterile H_2O, vortex briefly, spin down, and aspirate liquid.

14. Resuspend in 200 µL of YPD or TE and spread cells on appropriate selective plates (see Notes).

15. Incubate 2–3 d at the appropriate temperature for the strain.

    Notes:

    i. This protocol yields ~500–5000 colonies/µg of plasmid DNA. It is often wise to plate high and low volumes of cell suspension (e.g., 160 and 40 µL) to obtain plates with well-spaced transformants.

    ii. Integrating plasmids yield about 10–100-fold fewer transformants. A list of enzymes that digest the marker loci of pRS plasmids uniquely is provided below. Make sure that the enzymes do not digest your plasmid inserts.

    > *URA3*: *Stu*I, *Nco*I, *Nde*I (*Eco*RV if it has been deleted from polylinker)
    >
    > *LEU2*: *Xcm*I, *Afl*II, *Bst*EII at 60°C, *Age*I
    >
    > *TRP1*: *Bsu*36I, *Bst*Z17I, *Sna*BI, *Mfe*I
    >
    > *HIS3*: *Nhe*I, *Msc*I, *Nde*I
    >
    > *ADE2*: *Stu*I, *Hpa*I, *Afl*II, *Aat*II

    iii. To select for drug resistance markers (*KANMX, NATMX,* or *HYGMX*), first plate on YPD and then replica plate to YPD + drug the following day.

MATERIALS

1x LiOAc Buffer

0.1 M LiOAc
10 mM Tris-HCl (pH 8.0)
1 mM EDTA

> *Note:* It is convenient to make a 10× LiOAc stock from 1 M LiOAc and 10× TE.

1 M LiOAc

Dissolve 51 g of LiOAc (Sigma L6883-250G) in 450 mL of H_2O. Adjust the volume to 500 mL. Filter-sterilize or autoclave.

10x TE

10 mL of 0.5 M EDTA (Sigma 03690-100ML)
50 mL of 1 M Tris-HCl (pH 8.0; Sigma T2444-100ML)
440 mL of sterile ddH_2O

Single-Stranded Carrier DNA

10 mg/ml salmon sperm DNA (Applied Biosystems AM9680)
10 mM Tris-HCl (pH 8.0)
1 mM EDTA
Boil for 3 min, cool on ice, and store at −20°C in aliquots.

PEG Solution

Heat 50 mL of ddH_2O in microwave. To this add:

50 g of PEG 3350 (Sigma P3640)
10 mL of 10× TE
10 mL of 1 M LiOAc

Adjust the volume to 100 mL and filter-sterilize with a 0.45-μm filter unit (Nalgene). Store the PEG solution in a tightly capped container.

Using the C6 Cytometer

The C6 is a benchtop cytometer that has a lot of capabilities. For more detailed information, consult the user manual on the BD website: http://www.bdbio sciences.com (see cell analyzers section under instruments).

VOCABULARY

C6: The name of the flow cytometer.

CSampler: The silver arm attachment that allows multiple samples to be analyzed at once using either the tube holder or a multiwell plate.

CFlowPlus: The name of the software to use if you do *not* have the CSampler arm attached.

CFlowSampler: The software to use if you are using the CSampler attachment to analyze data (even if it is only one tube).

SIP: The needle that is used to sample your tubes. The SIP should not be left out of water with the power on. If you are going to be using the machine multiple times in a day and do not want to turn it off, make sure that the SIP is in distilled H_2O. To ensure that it remains in water, it is best to run it for 2 min with a 2-mL tube of water on the SIP, and then leave the tube there after the run has completed. *This is only to be done if the machine will be used again during the day. If you are the last user of the day, please shut down according to procedures on the laminated placard.*

Sheath Fluid: The fluid in the blue bottle. This fluid is what the machine runs through the inner tubing.

Decontamination Fluid: The fluid in the yellow bottle. This fluid is used automatically when the C6 is shutting down, or when you put the machine through a decontamination cycle. There is a Falcon tube of decontamination fluid near the C6 for use when shutting down.

Cleaning Fluid: The fluid in the green bottle. This fluid is used automatically when the C6 is shutting down. There is a Falcon tube of cleaning fluid near the C6 for use when shutting down.

Bead Validation: The C6 comes with two types of beads (6 peak and 8 peak) that can be used as validation that the machine is working properly. The BD representative already calibrated the machine when it was installed.

INSTRUCTIONS FOR USE

Always:

- Use MilliQ/Barnstead H_2O to make all solutions, and whenever running water through the machine. Do not use ddH_2O.

- Remember to leave the lid of the C6 closed while it is analyzing.

- Check the levels of the fluid bottles before starting your run. If there is not sufficient fluid, see the "maintenance" section below for instructions about making more.

- Make sure that you have the right software open. If the CSampler is attached, you want to open the CFlowSampler. Make sure the plate type that is selected in the software matches the plate type you are using. Also be sure that the plate is properly aligned in the robotic arm, with the "A1" corners lining up.

Start-Up

If someone used the machine before you, the needle should be sitting in a tube of water or the white "holding station," and you should start with Step 4. If you are the first user of the day, start with Step 1.

1. Check that the waste container is less than half full. Use the quick-disconnect valve above the bottle's cap to remove the bottle and dump the excess into the sink. Do not tug on the tubes! Press the metal tab to release. The tubes are attached to the back of the cytometer and *can be easily damaged,* so really—do not pull them taut.

2. Press the power button on the front of the machine and allow the lasers to warm up (5–10 min).

3. Open the *CSampler software* by double-clicking the icon on the desktop.

4. Get the 24-tube rack and put a 2-mL tube of *cleaning solution in A1,* and *H_2O in B1.* Open each tube, and tuck the caps into the adjacent tube spots.

5. Position the plate on the holder with A1 in the appropriate corner. Double check to make sure this is correct—it can cause a collision or the machine to

suck up air if the plate is not aligned correctly and the sipper goes to the wrong well.

6. Select the *Auto Collect* tab.

7. Go to the File menu, and select "Open CFlow file or template."

8. Select the appropriate cleaning file, which will end in .c6t (.c6 is a data file, and .c6t is a template file).

9. Click *Autorun*. It will run for 2 min in the cleaning solution, and then 2 min in the H_2O. The needle will go to its "parking spot," which is the white spot on the CSampler.

TIP! *The red box in the software that outlines one well of the plate at a time seems to indicate where the needle is, or is scheduled to go next. The check marks indicate the well selected for setting parameters.*

Running Your Samples

1. Click the *Auto Collect* tab, and *select the appropriate plate type*.

2. Go to the File menu, and select either
    a. "Open CFlow file or template"
        i. Select your template (.c6t file).

    b. Or "New CFlow file"
        i. Click on wells (individually or by rows/columns). A check mark will appear on selected wells.

        ii. Enter parameters for those wells.

        iii. Repeat as needed until all wells and parameters are loaded.

        iv. Under the File menu, select "Save CFlow template as" to name your template.

3. Name your plate, and click Open Run Display.

4. Then click Autorun. You will be prompted to name the file, so that your data will be automatically saved in the documents folder.

5. When the sampling is finished, click "close run display," then click on the Statistics tab.

6. On the left of the screen, select all your samples. At the top of the screen, click the boxes of the data types that you want to display.

7. Export your data using a USB drive:
   a. Under the File menu, you can export a .csv file, a plot file, etc. You can export to Excel and then e-mail or take the files with a flash drive.

   b. Alternatively, highlight the data in the table and paste it into a notepad document. The column and row labels will be transferred as well even though they are not highlighted.

8. Eject the plate, and dispose of your samples.

9. Immediately proceed to the Cleaning steps.

TIP! *Autosave only works if you are working from a template. If you work from a data file (.c6 instead of .c6t), the data collected will be cumulative, meaning it will add together every A1 count you ever ran with that file. Create a personal folder to keep all of your data in one place on the computer.*

After Sampling

1. Get the 24-tube rack and put a 2-mL tube of *cleaning solution in A1* and H_2O *in B1*. Open each tube and tuck the caps into the adjacent tube spots.

2. Position the plate on the holder with A1 in the appropriate corner.

3. Select the *Manual Collect* tab.

4. Go to the File menu, and select "Open CFlow file or template."

5. Select the appropriate .c6t file for cleaning.

6. Click on the A1 well and then click Run. It will run for 2 min.

7. Click on the B1 well and then click Run. It will run for 2 min, and the SIP (aka the needle) will remain in the water until you eject the plate.

8. If someone will be using the machine after you, you can leave the machine as soon as you start the water.

9. If you are the last user of the day, eject the tube rack, and press the power button on the front of the machine. It will go through a shutdown procedure that takes about 15 min. *Do not* hold the button down for longer than 1 sec. That sends it into emergency shutdown mode and causes complications for restarting it.

If You Need to Add a Solution

1. The software will warn you if you need to stop to refill a solution.

2. If you need to refill one, another one will likely run out soon (Sheath and Cleaning fluids are most used).

3. *Empty the waste bottle* (red) *every time you add sheath fluid!* Otherwise it will overflow.

4. Use the quick-disconnect valve above the waste bottle's cap to remove the bottle and dump it into the sink. Do not tug on the tubes! Press the metal tab to release them.

5. Mix the proper solution (see below) and add to the bottle. *Be sure to read directions*! Most of the solutions do not require the entire bottle of chemicals—they are concentrated.

6. Solutions are made from concentrate plus MilliQ H_2O.

7. The quick-disconnect valve will *click* when properly attached.

Sheath Fluid: Get 1 L of MilliQ H_2O, and add one bottle of concentrated bacteriostatic solution, mix, and add to the blue container.

Decontamination Fluid: Get 180 mL of MilliQ H_2O, and add one bottle of concentrated decontamination solution, mix, and add to the yellow bottle.

Cleaning Fluid: Dilute 3 mL of the concentrate in 197 mL of MilliQ H_2O.

Modified Hoffman–Winston Genomic DNA Preparation

SAFETY NOTES

Perform all steps with phenol in a chemical fume hood. Blue nitrile gloves are more phenol-resistant.

PROCEDURE

1. Grow an overnight culture in 2–5 mL of YPD or appropriate selective media.

2. Transfer 1.5 mL of culture to a 1.5-mL screw-top or lid-lock tube. This precaution prevents tubes from popping open during phenol extraction.

3. Pellet cells by centrifugation, decant, and remove residual liquid by aspiration. Repeat Step 2 if more cells are desired. Store the cell pellets at −80°C for later processing.

4. Add in the following order:

   a. 300 mg of acid-washed glass beads ~500-µm size range (~150 µL)

   b. 500 µL of lysis buffer

   c. 500 µL of 25:24:1 phenol:chloroform:isoamyl alcohol

   Note: Make sure that no glass beads interfere with the tube seal or contents will spill. Label tubes with permanent marker just in case. Black Sharpie is better than other colors.

5. Vortex vigorously for 5 min. Some vortexers are equipped with a multisample head that permits processing many samples simultaneously, although test these first to ensure adequate agitation.

6. Centrifuge for 5 min at maximum speed in a microcentrifuge.

7. Carefully transfer aqueous (top) layer to a new tube without catching the interphase that contains cell debris. Dispose the bottom layer in a liquid

phenol waste container and dispose the tubes and tips in a solid phenol waste container.

8. *Optional:* Extract the supernatant a second time with another 500 µL of phenol: chloroform:isoamyl alcohol, vortex, spin to separate phases, and transfer aqueous phase to a new tube.

9. Add 1 mL of room temperature 100% ethanol and invert to mix.

10. Centrifuge for 2 min at maximum speed. You should see a white pellet.

11. Resuspend the pellet in 50 µL H_2O.

> **Note:** This prep is sufficient for PCR analysis and transformation of yeast plasmids into bacteria. If you need a cleaner prep (for sequencing, microarrays, etc.) proceed to Step 12.

12. Resuspend pellet in 400 µL of 1× TE and add 30 µg of RNase A from a 10 mg/mL stock solution of TE. The pellet may be difficult to resuspend. If so, pipette up and down periodically during incubation in Step 13.

13. Incubate for 30 min at 37°C.

14. Add 10 µL of 4 M ammonium acetate and 1 mL of room temperature 100% ethanol. Invert to mix.

15. Centrifuge for 2 min at maximum speed in a microcentrifuge.

16. Remove supernatant completely and dry the pellet. This can be done by inverting the tube onto a KimWipe on the bench for 10–30 min.

17. Resuspend pellet in 50 µL of TE.

18. Measure DNA concentration using a fluorometer or other DNA-specific method (i.e., *not* the nanodrop). Even with the RNase treatment and ammonium acetate precipitation, there is a lot of RNA contamination in these preps. Total yield should be 10–20 µg. DNA should be suitable for a number of applications, including restriction digestion.

MATERIALS

Lysis Buffer

2% Triton X-100
1% SDS
100 mM NaCl
10 mM Tris-HCl (pH 8)
1 mM EDTA

Indirect Immunofluorescence Microscopy

SAFETY NOTES

Paraformaldehyde is highly toxic and volatile. It is also a possible carcinogen. It is readily absorbed through the skin and is irritating to the eyes, skin, mucous membranes, and upper respiratory tract. Wear gloves and safety glasses and always work in a chemical hood.

4′,6-Diamidino-2-phenylindole (DAPI) is a possible carcinogen. It may be harmful if it is inhaled, swallowed, or absorbed through the skin. It may also cause irritation. Wear gloves, facemask, and safety glasses, and do not breathe the dust.

The phenylenediamine in the mounting solution is a toxin and a potential carcinogen. Do not allow the mounting solution to contact the skin.

PROCEDURE

Cell Fixation (Done for You by the TA)

1. Grow 25 mL of cells to mid-log phase.

2. Transfer cells to a 50-mL conical tube and harvest by centrifugation (3–5 min by centrifugation at ~3000 rpm).

3. Resuspend in 1 mL of 4% paraformaldehyde solution.

4. Rock/rotate cells for 45–60 min at room temperature.

5. Pellet the cells and wash 2× in 5 mL of 0.1 M KPO_4 and 1× in 0.1 M KPO_4/1.2 M sorbitol.

6. Resuspend cells in 1 mL of 0.1 M KPO_4/1.2 M sorbitol.

7. Cells can be used immediately or stored at 4°C for several days.

Spheroplasting (Start Here on Day 3)

1. Transfer 0.5 mL of fixed cells to a 1.5-mL Eppendorf tube.

2. Add 10 µL of ß-mercaptoethanol (Sigma M6250-100ML).

3. Add 6 µL of 10 mg/mL zymolase 100T (US Biologicals Z1005).

4. Incubate cells until spheroplasted, generally 45 min. Monitor spheroplasting with a microscope—cells should be dark translucent gray when adequately spheroplasted. Overspheroplasted cells do not stain well as the antibody tends to bind to structures throughout the cell. Underspheroplasted cells also do not stain well as the antibody has trouble getting into the cell.

5. Centrifuge the cells for 1 min at 5000 rpm in a microcentrifuge.

6. Wash 2× with 1 mL of 0.1 M KPO$_4$/1.2 M sorbitol, resuspending cells gently with the blue tip for each wash (*do not vortex*).

7. Resuspend cells in 50–200 µL of 0.1 M KPO$_4$/1.2 M sorbitol.

8. Store the cells overnight at 4°C at this point.

Slide Preparation

1. Chill methanol in a coplin jar or other container at −20°C well ahead of time.

2. Fill another coplin jar with acetone and keep at room temperature.

3. Wash chamber slide (Polysciences Inc. 18357-1) with 95% ethanol and air dry.

4. Label the slide with a pencil (do not use pen, including sharpies as the ink will come off).

5. Place 15 µL of 0.1% polylysine (Sigma P8920-100 ml) into each well.

6. Let set for 5–10 min at room temperature.

7. Wash 3× with ddH$_2$O, then allow to air dry (this can be done with a dropper bottle or with a P200 pipetman).

8. Place 15 µL of spheroplasted cells into each well (make sure you have enough wells for controls, too).

9. Let set 5–10 min at room temperature.

10. Aspirate supernatant off cells.

11. Immediately plunge slide into ice-cold methanol for 5–6 min.

12. Remove slide and immediately submerge in acetone for 30 sec.

13. Quickly air dry the slide.

14. Immediately add 25 µL of PBS-BSA (phosphate-buffered saline–bovine serum albumin) to each well of slide using the dropper bottle or P200.

15. Do not let cells dry out until they are mounted.

Antibody Staining

1. While the cells are blocking with PBS-BSA (10–30 min), prepare primary antibody solutions by diluting antibody in PBS-BSA and centrifuging for 5 min at full speed in a microcentrifuge. This spin pellets particulates that will appear as fluorescent blobs in the microscope.

2. Aspirate off the PBS-BSA block.

3. Add 25 µL of solution of diluted primary antibodies to each well. This protocol assumes the use of two antibodies, one from rabbit and one from rat. Dilute each antibody according to specifications of the provider.

4. Incubate overnight at room temperature in a humid chamber (place lightly wet paper towels into a Tupperware container and seal).

CONTINUE HERE ON DAY 4

5. Prepare a solution of secondary antibodies in PBS-BSA and centrifuge for 5 min at full speed in a microcentrifuge. Dilute each secondary antibody according to the specifications of the provider.

   Goat α rat FITC (Jackson 112-095-003)

   Donkey α rabbit Cy3 (Jackson 711-165-152)

6. Aspirate off primary antibodies.

7. Wash wells 5× with PBS-BSA using the dropper bottle.

8. Add 25 µL of diluted secondary antibodies to each well.

9. Incubate 1–2 h in humid chamber in the dark at room temperature (put in drawer or cover with foil).

10. Aspirate off secondary antibody.

11. Wash wells 5× with PBS-BSA using the dropper bottle.

12. Wash wells 2× with PBS using the dropper bottle.

Optional DAPI Staining

1. Add 25 µL of DAPI (1 µg/mL DAPI in PBS).

2. Incubate 5 min at room temperature.

3. Aspirate off DAPI.

4. Wash 2× with PBS using the dropper bottle.

5. Let the slide air dry.

Mounting

1. Rinse coverslip with 95% ethanol, wipe with lens paper, and let air dry.

2. Use a stick or toothpick to add a dollop of mounting media (Citifluor by Ted Pella 19470) to the slide.

3. Put coverslip on top and gently push the coverslip to make sure the mount covers the entire slide and that all air bubbles are expelled.

4. Blot excess medium on a paper towel.

5. Seal edges with nail polish (Sally Hansen Hard as Nails).

6. When polish is dry, wash slide with H_2O and dry with a KimWipe.

7. Slides can be stored for several weeks at −20°C in the dark.

MATERIALS AND SOLUTIONS

4% Paraformaldehyde Solution

Dissolve 2.72 g of sucrose in 60 mL of H_2O.
Mix in 2 × 10 mL ampules of 16% paraformaldehyde from Ted Pella (break open and remove liquid with needle attached to 10-mL syringe; 18505).
Filter-sterilize.
Store at 4°C (this is good for 3–4 mo).

1 M KH_2PO_4

Dissolve 68 g in 500 mL of warm H_2O and then autoclave.

1 M K_2HPO_4

Dissolve 87 g in 500 mL of H_2O and then autoclave.

0.1 M KPO$_4$

41.7 mL of 1 M K$_2$HPO$_4$
8.3 mL of 1 M KH$_2$PO$_4$
450 mL of H$_2$O
Filter-sterilize.

0.1 M KPO$_4$/1.2 M Sorbitol

41.7 mL of 1 M K$_2$HPO$_4$
8.3 mL of 1 M KH$_2$PO$_4$
109.23 g of sorbitol
Dissolve and adjust volume with H$_2$O to 500 mL.
Filter-sterilize.

PBS-BSA

5 g of BSA (Sigma B4287-5G)
5 mL of 1 M KH$_2$PO$_4$
20 mL of 1 M K$_2$HPO$_4$
15 mL of 5 M NaCl
1 g of NaN$_3$
Dissolve and adjust volume with H$_2$O to 500 mL.
Filter-sterilize.
Store at 4°C.

PBS

5 mL of 1 M KH$_2$PO$_4$
20 mL of 1 M K$_2$HPO$_4$
15 mL of 5 M NaCl
1 g of NaN$_3$
Dissolve and adjust volume with H$_2$O to 500 mL.
Filter sterilize.

DAPI

Dissolve at 1 mg/mL in ddH$_2$O.
Centrifuge for 15 min at high speed in a microcentrifuge.
Remove supernatant and aliquot into dark-colored tubes.
Store at −20°C.

Secondary Antibodies

Dissolve in PBS at 1 mg/mL.

Centrifuge for 15 min at high speed in a microcentrifuge.

Remove supernatant and aliquot into dark-colored tubes.

Store at −20°C.

> *Note:* It may be necessary to titrate each new batch of secondary antibodies to determine the optimum working concentration.

Yeast Vital Stains

For a more extensive discussion of the use of vital dyes, see the methodology articles by Baggett et al. (2003) and Pringle et al. (1989).

A. DAPI STAINING OF NUCLEAR AND MITOCHONDRIAL DNA

Safety Notes

4′,6-Diamidino-2-phenylindole (DAPI) is a possible carcinogen. It may be harmful if inhaled, swallowed, or absorbed through the skin. It may also cause irritation. Wear gloves, a face-mask, and safety glasses and do not breathe the dust.

PROCEDURE

1. Briefly pellet ~10^7 cells in a microcentrifuge tube and resuspend in 70% ethanol. This step fixes cells for better DAPI staining.

   Note: 70% ethanol abolishes GFP fluorescence. See alternative procedures below for a formaldehyde fixation procedure that permits simultaneous visualization of GFP and DAPI.
   For staining live cells, see alternative DAPI procedures below.

2. Incubate for 5 min or more and wash twice with H_2O.

3. Resuspend cells in 50 μL of 50 ng/mL DAPI (Sigma D9542) in PBS. A stock of 1 mg/mL DAPI in H_2O can be stored at −20°C.

4. Observe with DAPI filter set.

Alternative Procedures

DAPI Staining of Live Cells

Add DAPI (final concentration ≅ 1 μg/mL) directly to the growth medium and incubate for 10 min. Staining is often weak, variable, and accompanied by higher background than it is with fixed cells. Staining can often be improved by adding Triton

X-100 (Sigma T8787) to the medium (final concentration = 0.1%). The nucleus can also be visualized in live cells by expressing a fluorescent protein fused either to a nuclear localization signal or to a nucleus-targeted protein.

DAPI Staining of Formaldehyde-Fixed Cells (Good for Preservation of GFP)

1. Pellet 1 mL of cells from mid-log culture.

2. Resuspend in 100 µL of paraformaldehyde solution. Incubate for 15 min.

3. Pellet cells. Wash once in KPO_4/sorbitol buffer. Cells can be stored in KPO_4/sorbitol buffer at 4°C for 1 mo.

4. Resuspend in 70% ethanol. Wait for 5 min or more before washing twice with H_2O.

5. Resuspend cell pellet in 25 µL of 50 ng/mL DAPI.

MATERIALS

Paraformaldehyde Solution

3.4 g of sucrose dissolved in 65 mL of H_2O.
1 mL of 1 M potassium phosphate buffer (pH 7.5).
25 mL of 16% paraformaldehyde (Ted Pella 18505). Add a few drops of 1 N NaOH if making from solid.
Adjust to 100 mL with H_2O. Filter-sterilize. Store at 4°C.

> *Note:* Paraformaldehyde is highly toxic and a possible carcinogen. Wear gloves and safety glasses and work in a chemical fume hood.

KPO_4/Sorbitol Buffer

60 mL of 2 M sorbitol
10 mL of 1 M potassium phosphate buffer (pH 7.5)
30 mL of H_2O

2 M Sorbitol

182 g per 500 mL of warm H_2O
Filter-sterilize.

1 M Potassium Phosphate Buffer (pH 7.5)

83.4 mL of 1 M K_2HPO_4
16.6 mL of 1 M KH_2PO_4

1 M K$_2$HPO$_4$

87 g per 500 mL of H$_2$O
Filter-sterilize.

1 M KH$_2$PO$_4$

68 g per 500 mL of H$_2$O warm
Filter-sterilize.

Use of mounting medium: Cells are sometimes mounted on slides in medium containing agents that retard photobleaching of fluorescent dyes and proteins (e.g., phenylenediamine or Citifluor). DAPI can be included in such mounting medium, or cells can be treated with DAPI, washed, and resuspended in mounting medium, as described here.

Mounting medium with DAPI: Dissolve 100 mg of *p*-phenylenediamine in 10 mL of PBS, adjust pH to above 8.0 with 0.5 M Na carbonate buffer (pH 9.0), and bring the volume to 100 mL with glycerol. Add DAPI to 50 ng/mL. Mix thoroughly and store at −20°C. It turns brown when it is bad.

> *Note:* Phenylenediamine is a toxin and a potential carcinogen. Do not allow contact with skin.

Mounting medium without DAPI: Incubate cells in 1 μg/mL DAPI for ~10 min, and then wash cells once with PBS. Place 5 μL of cell suspension on a glass slide and add 1–2 μL of Citifluor (Ted Pella 19470).

B. VISUALIZATION OF MITOCHONDRIA WITH MITOTRACKER, DIOC$_6$(3) OR DIIC$_5$(3)

To use MitoTracker Red (1 mM stock in DMSO, Life Technologies M7512), dilute $1/10^3$ to a final concentration of 100 nM and incubate for 30 min. Wash cells twice with SC, resuspend in 50 μL of SC and visualize with rhodamine filter set. MitoTracker must be used with live cells. However, once bound, the cells can be fixed with formaldehyde. This is a benefit over older dyes and is useful for colabeling experiments.

To use DiOC$_6$ (3,3′-dihexyloxacarbocyanine iodide; Sigma D3652), dilute $1/10^4$ from a 1 mg/mL stock in ethanol, incubate for 5–10 min, and observe with a fluorescein filter set.

To use DiIC$_5$(3) (1,1′-dipentyl-3,3,3′,3′-tetramethylindocarbocyanine iodide; Molecular Probes), dilute 1/25,000 from a 2.5 mg/mL stock in ethanol, incubate for 5–10 min, and observe with a rhodamine filter set.

> *Note:* The concentration of these dyes may need to be optimized. At high levels, mitochondrial specificity is lost and all membranes are stained.

C. VISUALIZATION OF VACUOLES AND ENDOCYTIC COMPARTMENTS WITH FM 4–64

1. Grow 5 mL of culture to mid-log.

2. Harvest 1 mL and resuspend in 200 μL of YPD.

3. Add 2 μL of 8 mM FM 4–64 (Molecular Probes T-3166; 1 mg in 200 μL of H_2O = 8 mM) and incubate for 30–60 min in the dark at 30°C. Longer incubations give more selective staining for membranes of acidic compartments.

4. Pellet cells and wash with 1 mL of YPD.

5. Resuspend in 1 mL of YPD and grow for an additional 60 min at 30°C.

6. Centrifuge and resuspend in 50 μL of YPD. Examine using the rhodamine filter set.

D. CALCOFLUOR STAINING OF CHITIN AND BUD SCARS

1. To ~10^7 cells in growth medium, add calcofluor (Fluorescent Brightener 28; Sigma F3543) to a final concentration of 100 μg/mL (dilute 1/10 from 1 mg/mL stock in H_2O, which is stable for weeks in the dark at −20°C).

2. Incubate for 5 min or more, wash twice with H_2O and resuspend in 50 μL. Observe with a DAPI compatible filter set.

REFERENCES

Baggett JJ, Shaw JD, Sciambi CJ, Watson HA, Wendland B. 2003. Fluorescent labeling of yeast. In *Current protocols in cell biology* (ed. Bonifacino et al.), Chapter 4: Unit 4 13. John Wiley and sons, Hoboken, New Jersey.

Pringle JR, Preston RA, Adams AEM, Stearns T, Drubin DG, Haarer BK, Jones EW. 1989. Fluorescence microscopy methods for yeast. *Methods Cell Biol* **31:** 357–435.

Actin Staining in Fixed Yeast Cells

SAFETY NOTES

Paraformaldehyde is highly toxic and volatile. It is also a possible carcinogen. Always use in a chemical fume hood. Avoid breathing vapors and wear appropriate gloves and safety glasses. Keep away from heat, sparks, and open flame.

PROCEDURE

The following eight steps were performed by the TA:

1. Grow a 25-mL culture of yeast in YPDA to mid-log phase at the appropriate temperature.

2. Pour media into a 50-mL conical tube containing 10 mL of 16% paraformaldehyde from Ted Pella (break open ampules and remove liquid with needle attached to a 10-mL syringe in a chemical fume hood).

3. Incubate for 10 min with gentle shaking.

4. Pellet cells by centrifugation at 2000–3000 rpm for 3–5 min.

5. Collect supernatant in a chemical fume hood.

    Note: Formaldehyde waste should be neutralized before disposal by adding glycine (final concentration = 125 mM using a 2.5 M stock solution) or by adding Neutralex® (see manufacturer's instructions).

6. Resuspend cell pellet in 4% formaldehyde in PBS. Incubate for 1 h at room temperature.

7. Wash cells twice with 10 mL of PBS and resuspend in 500 µL of PBS in a 1.5-mL Eppendorf tube. Dispose of formaldehyde solutions as described above.

8. Store at 4°C overnight. (It is possible to proceed immediately to Step 9, but this is a convenient stopping point.)

Students Begin Protocol at Step 9

9. Add 20 μL of rhodamine-phalloidin or fluorescein-phalloidin (the stock solution is 6.6 μM in methanol) to 180 μL of cell suspension. Incubate for 1 h in the dark. Vortex approximately every 15 min.

10. Wash cells five times in 1 mL of PBS.

11. Resuspend the cells in 1 mL of PBS and then add 1 μL of 1 mg/ml DAPI and incubate at room temperature for 5 min.

12. Pellet/rinse twice with PBS.

13. Resuspend cells in 200 μL of PBS.

14. Visualize the cells by placing 8 μL of the cell suspension on a glass slide and covering with a coverslip.

MATERIALS AND SOLUTIONS

PBS

8 g of NaCl
0.2 g of KCl
1.44 g of Na_2HPO_4
0.24 g of KH_2PO_4

Dissolve in 1 L total volume, adjust pH to 7.2, filter-sterilize, and store at 4°C.

16% Paraformaldehyde (Ted Pella 18505)

Rhodamine-phalloidin (Molecular Probes R-415)

Dissolve in methanol according to the manufacturer's instructions to ~6.6 μM, aliquot, and store at −20°C.

DAPI (Sigma D9542)

Dissolve at 1 mg/mL in ddH$_2$O, aliquot, and store at −20°C.

Preparation of Slides with Agarose Pads for Imaging of Live Yet Immobile Yeast

PROCEDURE

1. *Preparation of stock aliquots of agarose:* Add 0.26 g of ultrapure agarose to 20 mL of SC media containing 4% dextrose. Microwave until agarose dissolves and dispense 500-μL aliquots into Eppendorf tubes. Store at 4°C.

2. *Preparation of agarose pads:* Melt a single aliquot of agarose solution in a 95°C temperature block. Clean a depression well slide (Fisher 50-949-458) with ethanol. Pipette 160 μL of agarose solution into well and slide a standard microscope slide on top of the well. Avoid bubbles. Allow agarose to solidify (15 min maximum).

3. *Preparation of cells:* While agarose is solidifying, transfer 1 mL of cells from a mid-log phase culture to an Eppendorf tube and centrifuge at 6500 rpm for 2 min in a microcentrifuge. Sometimes the pellet is difficult to see because the cells spread along the side of the tube. If so, carefully remove half of the volume and respin. Repeat until a pellet forms at the bottom of the tube. Resuspend in 20–30 μL of media. Synthetic media is preferable because YPD is autofluorescent.

4. *Final preparation of slide:* Remove the standard microscope slide from the well slide with one quick sideways shearing motion (Rines et al. 2011). Remove residual fragments of agar that will prevent the coverslip from laying flat using the edge of the standard microscope slide.

5. *Addition of cells:* Spot 3 μL of concentrated cell culture on the surface of agarose. Adding too much volume causes cells to float. Gently place a 22 × 22-mm coverslip on the liquid. Do not move the coverslip once in place or the cells will move and crowd one another. Seal the edges with nail polish to slow evaporation and prevent movement.

Note: Some commercial preparations of yeast nitrogen base are fluorescent and yield a high background in agarose pads. If this is a problem, the nitrogen base can be assembled from pure components (see Sheff and Thorn 2004).

REFERENCES

Rines DR, Thomann D, Dorn JF, Goodwin P, Sorger PK. 2011. Live cell imaging of yeast. *Cold Spring Harbor Protocols*. PMID: 21880825.

Sheff MA, Thorn KS. 2004. Optimized cassettes for fluorescent protein tagging in *Saccharomyces cerevisiae*. *Yeast* **21:** 661–670.

Sporulation and Tetrad Dissection

Unfortunately, there is no single magic sporulation protocol for all yeast strains. Some strain backgrounds such as SK1 sporulate efficiently and rapidly even when grown on rich media, whereas other strain backgrounds such as BY, which was used to construct the yeast deletion collection, sporulate slowly and inefficiently. For a comprehensive study of sporulation in commonly used yeast strains, see Elrod et al. (2009).

Below is a general liquid protocol that is useful for sporulation of BY-, S288C-, and W303-derived strains. A plate-based sporulation protocol can be used as an alternative. Both methods should yield asci with four spores in 3–5 d at 23°C, although BY strains require longer times for optimal sporulation (∼5–7 d). Following digestion with zymolyase, which is also described below, tetrads can be dissected.

LIQUID SPORULATION PROTOCOL

1. Grow diploid strain in 2 mL of overnight culture in Pre-SPO (or YPD) media at 23°C.

2. Centrifuge the cells from 1 mL of overnight culture and wash 2× with SPO media or sterile H_2O.

3. Resuspend cells in 0.5 mL of SPO media.

4. Transfer to a sterile culture tube containing 2 mL of SPO media.

   Note: Culture density is important for efficient sporulation. Cell suspension should be visibly cloudy but not opaque. An OD between 0.1 and 0.3 generally works well. Also, it is important for the culture tube to contain no more than 4 mL of culture. Too much liquid will prevent efficient mixing.

5. Incubate SPO culture in a shaking incubator at 23–25°C overnight.

6. Shift SPO cultures to a 30°C shaking incubator *unless* your strains contain temperature-sensitive mutations.

   Note: Many investigators find that this produces a more robust sporulation in S288C. However, some mutants, including alleles that are *not* temperature-sensitive, do not respond well to the temperature shift. Often, it is useful to divide the culture in half, keeping 2 mL at 23°C and shifting 2 mL to 30°C.

7. Usually, cultures need ~3–4 d before they have sporulated enough to dissect. It is possible to continue sporulation for up to 2 wk. However, this will result in a thicker ascus wall.

8. Sporulated cultures can be stored at 4°C for up to 1 mo without significant loss of spore viability.

Alternative Media for Plasmids

If you are trying to retain a plasmid, use dropout media supplemented with 4% glucose for the preculture described in Step 1 above. Other investigators prefer taking diploids from a very *fresh* plate. Either way, some diploids will lose the plasmid and you will need to dissect a few more plates to ensure you get your strain.

PLATE SPORULATION PROTOCOL

1. Streak the diploid strain for single colonies on YPD or a selective dropout plate to maintain plasmids. Grow at appropriate temperature until single colonies are visible.

2. Pick a single colony and patch a thin layer of cells to one quarter of a SPO plate.

3. Incubate the SPO plate at 23°C for 3–5 d until spores are visible. It is possible to continue sporulation for up to 2 wk. However, this will result in a thicker ascus wall.

4. Plates with sporulated cells can be stored at 4°C for up to 1 mo without significant loss of spore viability.

PREPARATION OF TETRADS

1. Sporulate cells on either plates or in liquid medium using the protocols described above. Examine the sporulated cultures to confirm that tetrads have been produced. For liquid cultures, put 5 µL of the sporulation mix on a slide. For cells that have sporulated on plates, put 5 µL of H_2O on the slide and then use a toothpick to pick up a clump of cells from the SPO plate and suspend in the H_2O. Add coverslips and visualize on the compound microscope. Ideally, we would like to achieve at least 30%–40% of cells containing four-spored tetrads. Cultures showing less than 5% tetrads are difficult to dissect.

2. Prepare a fresh zymolyase cocktail for digestion of the ascus coat. Zymolyase contains β-glucuronidase, which cleaves bonds in the ascus coat, making it

easier to break apart the ascospores. A less expensive alternative to zymolyase is Glusulase (typically used as a 1:10 dilution of the stock), which generally yields less efficient digestion of the ascus coat, with greater spheroplasting of ascospores, than does zymolyase. We will add sorbitol as an osmotic stabilizer and our cocktail will also contain ß-mercaptoethanol, which reduces disulfide bonds in the cell wall. For each sample, combine

200 µL of 0.1 M KPO$_4$/1.2 M sorbitol solution

10 µL of ß-mercaptoethanol (Sigma M3148-100ML)

10 µL of zymolyase 100 T (US Biologicals Z1005, 10 mg/mL)

3. If sporulation was performed in liquid, add 1 mL of the culture to a microcentrifuge tube, centrifuge at 5000g for 10 sec, and remove the supernatant. Gently resuspend the cell pellet in 200 µL of zymolyase cocktail. If the sample was sporulated on plates, use the flat end of a sterile toothpick to transfer a dab of cells from the plate to a microcentrifuge tube containing 200 µL of the zymolyase cocktail. Suspend the cells by rotating the toothpick. Incubate at 23°C on your bench.

4. Stop the zymolyase digestion by placing the microcentrifuge tube on ice and *gently* adding 200 µL of 0.1 M KPO$_4$/1.2 M sorbitol solution. For most strains, an incubation time of ~10 min is appropriate, but some strains require significantly longer or shorter incubations. The appropriate incubation time can also vary for a specific strain as a function of sporulation time and conditions (e.g., liquid vs. solid sporulation medium). If you have not previously dissected your strain, a good strategy is to remove samples from the zymolyase digestion at timed intervals (2–20 min). These samples can be examined microscopically to determine the ideal incubation time. Optimal digestion will convert a large fraction of the asci from tight pyramid-shaped structures to looser diamond-shaped structures (see micrographs in Experiment IV). Overdigestion breaks open the diamond-shaped structures and releases ascospores.

5. Use a Sharpie to mark a YPD plate where cells will be placed. Generally, the stripe of zymolyase-treated cells is made across the top or center of the plate (in the center if you are dissecting tetrads on the SporePlay). The cells must be accessible to the micromanipulator, and there must be sufficient space either above or below the stripe for placement of isolated ascospores.

6. Very gently apply 10 µL of zymolyase-treated cells as a streak across a YPD plate using a sterile loop. Alternatively, use a pipettor to transfer the cells to a tilted plate, allowing the droplet of cell suspension to run down the sloped agar surface,

leaving a stripe of zymolyase-treated cells. With either method, care must be taken not to disrupt the tetrads that are now only tenuously intact because of the zymolyase treatment. Wait several minutes for the liquid to absorb into the plate, which can then be mounted on a suitable tetrad dissection microscope.

MATERIALS

Pre-SPO Media, and Minimal SPO Media or SPO Plates

See Appendix A

1 M KH_2PO_4

Dissolve 68 g in 500 mL of warm H_2O and then autoclave.

1 M K_2HPO_4

Dissolve 87 g in 500 mL of H_2O and then autoclave.

0.1 M KPO_4

41.7 mL of 1 M K_2HPO_4
8.3 mL of 1 M KH_2PO_4
450 mL of H_2O
Filter-sterilize.

0.1 M KPO_4/1.2 M Sorbitol

41.7 mL of 1 M K_2HPO_4
8.3 mL of 1 M KH_2PO_4
109.23 g of sorbitol
H_2O to 500 mL
Filter-sterilize.

Zymolyase Cocktail

200 µL of 0.1 M KPO_4/1.2 M sorbitol solution
10 µL of ß-mercaptoethanol (Sigma M3148-100ML)
10 µL of zymolyase 100T (US Biologicals Z1005, 10 mg/mL)

REFERENCE

Elrod SL, Chen SM, Schwartz K, Shuster EO. 2009. Optimizing sporulation conditions for different *Saccharomyces cerevisiae* strain backgrounds. *Methods Mol Biol* **557**: 21–26.

Making a Tetrad Dissection Needle

Dissection needles, typically made from glass or optical fiber and attached to any of a variety of types of micromanipulators, are useful in the dissection of tetrads, isolation of zygotes from populations of mating haploid cells, and manipulation of individual cells. For the Singer SporePlay dissection microscopes that we use in the course, needles are purchased from Singer premounted on a glass holder (http://www.singerinstruments.com). For other types of micromanipulators, needles can be purchased from commercial sources (Cora Styles Lab Supplies, http://www.corastyles.com). Alternatively, it is possible to make needles by gluing a thin glass filament with a flat end to a bent glass capillary pipette.

Needles can also be made by drawing thin filaments from a 2-mm-diameter glass rod using a small gas flame. The exact diameter is not critical, and investigators have different preferences. Spores are more readily picked up and transferred with microneedles having tips of larger diameters (100 µm), whereas manipulations in crowded areas having high densities of cells are more manageable with microneedles having smaller diameters (25 µm). A needle with a diameter of ~50 µm is an acceptable compromise. To draw out the needle, hold the glass rod over a low Bunsen burner flame until it is glowing orange. Simultaneously, remove the rod from the flame while pulling the ends apart. With practice, it is possible to produce a glass thread using this method. Segments that appear to be the correct diameter are chopped into lengths of ~2 cm, placed on a glass slide that has been prewetted with water, and then cut with a razor blade or glass coverslip to create microneedles 1 cm in length. The goal is to create microneedles with one perfectly smooth flat end. The short segments are inspected with a microscope to determine whether they have the correct diameter and a flat end.

An alternative to the glass needle is a fiber-optic needle. Fiber-optic filament of ~50 µm can be purchased and cut into 1-cm needles with a razor blade. Like the glass microneedles, these can be inspected with a microscope to determine if the cut generated a flat surface. Next, the needle is attached to a 2-mm rod. Although it is possible to use 100-µL capillary pipettes for this purpose, metal rods are a more durable alternative. If using a capillary pipette, heat the glass rod over a low flame about 1 cm from its end, and when it becomes pliable, bend it to a right angle.

Bend the metal rod to form a right angle. A drop of Super Glue is applied to this end of the mounting rod, which is then touched to one of the segments of a glass filament. The filament is carefully positioned so that it is at a right angle to the axis of the mounting rod and it has the correct length. The length of the perpendicular end should be compatible with the distance between the needle holder of the micromanipulator being used and the surface of the dissection plate. A too-short needle will not reach the surface of the medium, and one that is too long will dig into the surface of the medium. The needle holder is fitted into the micromanipulator after the glue has dried. If a needle is broken, the Super Glue can be easily removed from the needle holder using fine-grit sandpaper, and a new needle can be mounted.

EMS Mutagenesis

SAFETY NOTES

EMS is a volatile organic solvent that is a mutagen and carcinogen. It is harmful if inhaled, ingested, or absorbed through the skin. Wear gloves and protective clothing, and work in a chemical fume hood when performing the mutagenesis. Use disposable tubes and plastic pipettes for all manipulations. There are waste containers containing 5% sodium thiosulfate located in the hood for both solid and liquid EMS waste. EMS should be stored in the refrigerator until immediately before use to minimize its volatility. Any glassware that comes in contact with EMS should be immersed in a large beaker of 1 N NaOH or laboratory bleach before recycling or disposal.

PROCEDURE

1. Grow 5 mL of overnight culture in YPD to saturation.

2. Transfer 1 mL of the overnight culture to two sterile 1.5-mL microcentrifuge tubes. Pellet the cells by centrifuging for 10 sec at 5000g in a microcentrifuge.

3. Wash the cell pellets twice with 1 mL of sterile ddH$_2$O.

4. Resuspend cells in 1 mL of sterile 0.1 M sodium phosphate buffer (pH 7.0).

5. Remove 10 µL of cells to determine the exact cell density, either during Step 6 or later.

6. To one of the two 990 µL of cell suspensions, add 30 µL of ethylmethanesulfonate (EMS). (The other tube will be the unmutagenized control.) Close the tops of both tubes with Parafilm and vortex vigorously to disperse the EMS. Incubate both tubes for 1 h at 30°C with occasional agitation.

7. While the cells are mutagenizing, use the cell aliquot from Step 5 to determine the exact cell density by counting with a hemocytometer. In addition, label tubes and plates required for cell dilutions and plating.

8. Pellet the mutagenized cells and nonmutagenized controls. Transfer the supernatant to a designated EMS waste container and then resuspend the cells in

200 μL of 5% sodium thiosulfate. Transfer cell suspensions to fresh tubes. Discard the used tubes in an EMS waste container.

9. Wash the cell pellets twice with 200 μL of 5% sodium thiosulfate (each time discarding the supernatant in an EMS waste container). Resuspend the cells in 1 mL of sterile ddH$_2$O.

10. Continue with the protocol in Experiment V. Some cells will be plated directly and others will be given a period of outgrowth. This allows wild-type proteins to be replaced by their mutated counterparts before exposing the cells to selective conditions. If outgrowth is performed, it is important to remember that this treatment results in the production of identical siblings in the culture.

MATERIALS AND SOLUTIONS

0.1 M sodium phosphate (pH 7.0)
    99 mL of 0.2 M sodium phosphate monobasic
    201 mL of 0.2 M sodium phosphate dibasic
    300 mL of dH$_2$O

0.2 M sodium phosphate monobasic
    13.9 g of sodium phosphate monobasic (Sigma S6566-100G)
    500 mL of dH$_2$O

0.2 M sodium phosphate dibasic
    53.65 g of sodium phosphate dibasic heptahydrate (S9390-100G; or 28.4 g of the anhydrous form S3264-250G)
    1 L of dH$_2$O

5% sodium thiosulfate
    5 g of sodium thiosulfate (Sigma S8503-500G)
    100 mL of ddH$_2$O
    Filter through a 0.2-μm filter, and dispense in 10-mL aliquots.

Counting Yeast Cells with a Hemocytometer

The optical density of a yeast culture measured in a spectrophotometer provides only a rough estimate of the number of cells present in the culture because it is greatly affected by cell size, growth media, strain background, and a number of other factors. The most accurate way to determine the number of cells is to use a standard hemocytometer. The surface of this specialized microscope slide contains a chamber that is etched with a grid pattern of lines (shown on next page). A rigid coverslip rests on top of risers adjacent to the chamber to create defined volumes demarcated by the etched grid lines. Cell density is determined by counting the number of cells within these defined volumes.

Clean the hemocytometer and special coverslip with ethanol and allow to air dry or blot with lens paper. Position the coverslip over the counting chamber such that it rests on the risers. Apply the sample so that it fills the counting chamber by capillary action. This can be done by pipetting 10 µL of liquid slowly into the groove at the bottom of the counting chamber until the area under the coverslip is filled. Do not overfill.

Use a 10× air objective to look at the cells in the chamber under the microscope. Find the large center square (circled in the figure) that contains 25 small squares (one of which is filled in with black in the figure), each of which is itself made of 16 tiny squares. Using a 40× objective, only a portion of the large square is visible, but it will be easier to determine the number of yeast cells in the small squares.

Count the cells in the entire large square. It is easiest to systematically count one small square at a time to keep track of your place. To avoid overcounting cells that fall on the borders, only count cells that cross the top and right borders, and not cells crossing the bottom and left borders (or whatever pair of borders is easiest for you to see). Some of the borders are made of three lines. The central line is the important one that gives the measured volume. Budding cells also present a challenge. Cells with small buds should obviously be counted as single cells. More ambiguous are two large adjacent cells. This could be a large bud on a cell that is still dividing, or it could be two separate cells. Use your judgment, and be consistent.

Sometimes you can see a hint of a small bud emerging from one cell, which allows you to definitively count them as two cells. If you see many clumps, you may need to try again with a new sample that has been more aggressively resuspended or even sonicated.

The coverslip is precisely 0.1 mm above the grid, and the large square is 1 mm square. Therefore, the volume of the entire large square is 0.1 µL and you can easily calculate the number of cells/mL in your sample by multiplying by 10^4.

In typical use, your culture is not generally going to be at a convenient density. You always want to count at least 100 cells to ensure accurate sampling. However, if the culture is too dense, you will not be able to accurately count the cells as they crowd together. Cells will usually need to be diluted so that an accurate count can be achieved. This can be done in H_2O or in media. Below is a general guide to typical cell counts for yeast cultures:

Culture	OD_{600}	Cell Count/mL	Dilution
Saturated YPD	25	3×10^8	10^{-2} or 10^{-3}
Log YPD	0.25	3×10^6	None or 10^{-1}
Saturated SC	5	1×10^8	10^{-1} or 10^{-2}

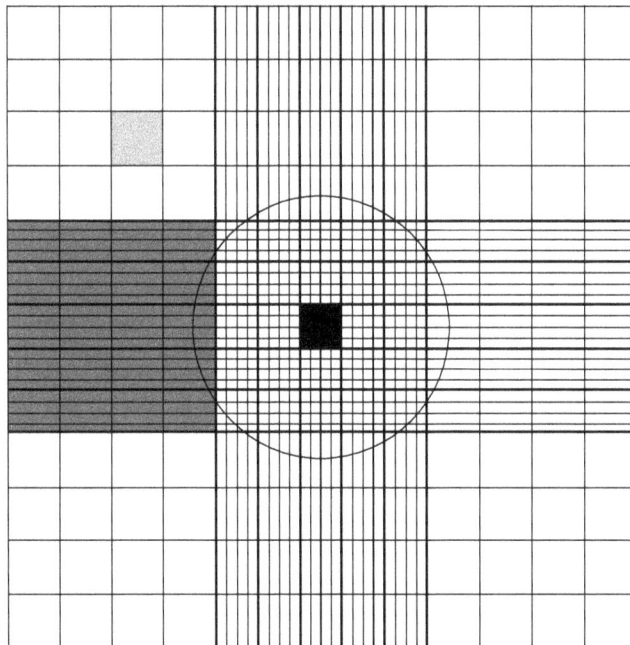

You also do not need to count all 25 of the small squares if you can get >100 cells in fewer compartments. For example, you could choose five small squares and count the number of cells present in just those squares. Use the same rules as above with regard to border-crossing cells, buds, etc. Also, do not just stop when you hit 100. You have to count the entire small square so you know the volume. For five small squares, the calculation would be

cells in 5 small squares $\times$ 5 $\times$ 10^4 $\times$ dilution factor = # cells/mL in your culture

The volume of each small square is 0.004 µL, and thus you can calculate for however many squares you like. Do not count less than five, however, to obtain a good sample.

Flow Cytometry of Yeast DNA Content Using SYBR Green

PROCEDURE

1. Remove 1 mL of cells from liquid culture at 0.5–1.0 OD_{600} and put into an Eppendorf tube (make sure to always include an asynchronous wild-type haploid culture to calibrate the flow cytometer).

2. Centrifuge the cells in a microcentrifuge at 14,000 rpm for 1 min.

3. Aspirate the supernatant (be careful not to suck up pellet!).

4. Add 1 mL of 70% ethanol and resuspend the pellet by vortexing.

5. Allow the cells to sit at room temperature for at least 1 h, and store them at 4°C or continue with the protocol.

6. Centrifuge 0.5 mL of cells and carefully aspirate the supernatant. Keep the other 0.5 mL of ethanol-fixed cells at 4°C.

7. Resuspend the cells in 0.5 mL of ddH_2O.

8. Centrifuge the cells and aspirate the supernatant (the pellet may not be visible, so be careful when aspirating).

9. Resuspend the cells in 200 µL of RNase A solution (4 µL of 10 mg/ml RNase A + 196 µL of 50 mM Tris-Cl (pH 8.0); make up a master mix for all your samples).

10. Incubate the cells at 37°C for 2–4 h.

11. Centrifuge the cells and aspirate the supernatant.

12. Resuspend the pellet in 200 µL of proteinase K solution (2 mg/mL proteinase K in 50 mM Tris-Cl [pH 7.5]; make up a master mix for all your samples).

13. Incubate at 50°C for 30–60 min.

14. Centrifuge the cells and aspirate the supernatant.

15. Resuspend in 200–400 µL of FACS buffer (store samples in FACS buffer at 4°C, but no longer than 1 wk).

16. Transfer 10 µL of cells into a 96-well plate. Add 200 µL of SYBRGreen (diluted 5000× in 50 mM Tris-Cl at pH 7.5) to each well. Samples are light-sensitive; keep in dark place where possible.

17. Sonicate each sample for 3 sec at 10% power (5–10 Watt output) to reduce clumping of cells.

18. Run samples immediately on the flow cytometer (see Techniques and Protocols 4).

MATERIALS

70% Ethanol (250 mL)

Mix 184.2 mL of 95% ethanol with 65.8 mL of H_2O.
Filter-sterilize.

1 M Tris-Cl (pH 7.5/8.0; 500 mL Each)

Add 60.57 g of Tris to 300 mL of H_2O.
Add HCl until solution reaches appropriate pH.
Adjust volume to 500 mL with H_2O.
Autoclave.

5 M NaCl (500 mL)

Add 146.1 g of NaCl to 300 mL of H_2O.
Adjust solution to near 500 mL and heat to dissolve.
Adjust volume to 500 mL with H_2O.
Autoclave.

1 M MgCl$_2$ (100 mL)

Add 20.33 g of $MgCl_2$ to H_2O.
When dissolved, adjust to 100 mL with H_2O.
Autoclave.

Proteinase K Solution

2 mg/mL proteinase K (BioShop PRK403) dissolved in 50 mM Tris (pH 7.5).
Make fresh.

RNase A Solution

10 mg/mL RNase A (Sigma-Aldrich R4642) stored at −20°C.

50 mM Tris-Cl (pH 7.5/8.0; 250 mL)

Add 12.5 mL of 1 M stock to 237.5 mL of H_2O.
Filter-sterilize.

SYBR Green

Stock solution from Sigma-Aldrich (S9430). Dilute 5000× in 50 mM Tris-Cl (pH 7.5) for working solution.

FACS Buffer (100 mL)

200 mM Tris-Cl (pH 7.5)	20 mL of 1 M stock
200 mM NaCl	4 mL of 5 M stock
78 mM $MgCl_2$	7.8 mL of 1 M stock
H_2O	68.2 mL

Storage and Handling of the Systematic Deletion Collection

Genetic and genomic analyses of budding yeast have developed dramatically over the past decade, in part due to publicly available collections of modified yeast strains for a variety of uses. There are collections of deletion mutants where almost every gene in the genome has been deleted by a one-step gene replacement with a *kanMX4* module. Conveniently, there are 20-bp unique oligonucleotide sequences that have been inserted with the *kanMX4* module that serve as unique identifiers and are called "barcodes." The unique barcodes can be amplified with a universal set of primers that amplify the 5′ barcode (uptag) and the 3′ barcode (downtag). The construction of the deletions is described in the *Saccharomyces* Deletion Consortium webpage (http://www-sequence.stanford.edu/group/yeast_dele tion_project/deletions3.html). There are other large collections of yeast strains where almost every open reading frame (ORF) is tagged with glutathione-*S*-transferase (GST) (Zhu et al. 2001; Sopko et al. 2006), green fluorescent protein (GFP) (Huh et al. 2003), or the tandem affinity purification (TAP) epitope (Ghaem-maghami et al. 2003). Comprehensive descriptions of common yeast strains, collections, and sources can be found in the *Saccharomyces* Genome Database wiki (http://wiki.yeastgenome.org/index.php/Strains). The deletion collection strains are available individually or *en masse* from vendors such as GE Healthcare Dharmacon (http://dharmacon.gelifesciences.com/non-mammalian-cdna-and-orf/yeast-knockout-collection/), transOMIC technologies (http://www.transomic.com/Prod ucts/Yeast-Products/Yeast-Knock-Out-Collection/Product-Overview.aspx), ATCC (http://www.atcc.org/Products/Cells_and_Microorganisms/Fungi_and_Yeast/Sac charomyces_cerevisiae_Deletion_Mutants.aspx), and EUROSCARF (http://www. euroscarf.de).

The deletion collection is available in three strain backgrounds that are isogenic with BY4741 (*MAT*a), BY4742 (*MAT*α), and BY4743 (*MAT*a/*MAT*α) (Brachmann et al. 1998). Mutants that have deletions of nonessential genes are available either as haploids or as heterozygous or homozygous diploids. If you are buying an individual strain, purchase the heterozygous diploid. Mutants with deletions of essential genes are only available as heterozygous diploids. If you are interested in a mutant with a deletion of a nonessential gene, you must dissect tetrads to obtain the mutant of the desired haploid genotype, isogenic with BY4741 or BY4742. Once the strain is obtained, it should be immediately frozen for further use. Deletion mutants are often compromised for growth, and prolonged passaging of haploid mutants can select for suppressors, which range from single point mutations to whole-chromosome aneuploidy. The reduced fitness of the deletion mutant is often recessive and therefore the problem of changing genotype is minimized by using diploids. Recovering haploids after meiosis reduces the number of times the haploids have been passaged. In addition, we recommend that you test for any other phenotype of your mutant to be sure that you have the right isolate and to confirm that it cosegregates with the deletion allele. The G418 resistance is not sufficient, as all of the deletion mutants are thus marked. We suggest that you amplify the barcodes using the strategy described by Ooi et al. (2001). The resulting "uptag" and "downtag" DNA fragments are then sequenced using the Sanger method (see Experiment IX). The sequences of the barcodes can be obtained from the supplementary material of Ooi et al. (2001), with additional error correction (Eason et al. 2004; Smith et al. 2009). This practice assures that you are working with the correct mutant.

Sometimes, it is necessary to purchase an entire collection. For example, Synthetic Genetic Array (SGA) analysis requires the entire deletion collection. The collection is sold in an ordered 96-well format. The collection shipped from the ATCC comes in 74 separate plates. The strains are shipped frozen in 150-µL glycerol stocks in sealed plates. The major concern in retrieving strains from the plates is cross contamination, and great care must be taken to prevent this. Thaw the plates at room temperature and centrifuge them for 1 min to remove any liquid that may be on the seal. Carefully remove the seal and transfer the strains from the original plate using a sterile 96-pin replica pinning tool, available from V&P Scientific (http://www.vp-scientific.com/). Once the strains are transferred, carefully reseal the plates with fresh seals (such as radiation sterilized Nalgene Nunc 96-well seals, 236366) and refreeze the plates at −80°C covered by the plastic lids. Use storage racks designed for microtiter plates. Create a database that indicates where the plates are stored. In subsequent use, remove the foil immediately upon taking the plates from the freezer, before thawing them at room temperature. If condensation develops on the plastic lid during thawing, wipe it off with an ethanol-dampened KimWipe.

The strains can be transferred to liquid media in 96-well plates or to solid agar media in OmniTrays (Nunc) or PlusPlates (Singer Instruments). Higher-density arrays are constructed by using a colony copier template (V&P Scientific) to correctly align four 96-colony arrays into a single 384-colony array and four 384-colony arrays into a single 1536-colony array. As an alternative to replica pinning by hand, automated systems are available to transfer colonies robotically (Singer Instruments [http://www.singerinstruments.com], S&P Robotics [http://www.sprobotics.com]). Propagation on media containing G418 reduces the risk of contamination.

To store a fresh copy of the deletion collection, the arrayed colonies should be pinned to a 384-well microtiter plate containing 50 μL of YPD plus 200 μg/mL G418. After 2 d growth at 30°C, add 25 μL of 45% glycerol, freeze at −80°C, seal with foil, and store using racks designed for microtiter plates. Make sure to add the new plates to your database. If possible, the collection should be stored in duplicate in separate freezers. One strain management technique is to designate a master copy of the collection that is not in frequent use so that freeze–thaw cycles and other handling are kept to a minimum. The second copy is a working collection that is used for experiments. Once the working collection has become untrustworthy (>10 freeze–thaw cycles, irregular volume in the wells, evidence of contamination, etc.), a new working copy can be generated from the pristine master copy.

There are two internal checks for the integrity of the collection. The first is that the original 96-well microtiter plates have some wells that contain no yeast strains. They are an excellent indicator of the extent of cross-contamination. If there is no growth in any of the wells that should be empty, then cross-contamination is minimal. The second test is the phenotype of large classes of mutants. Pin 384 plates to OmniTrays containing YPD, YPG, and SD agar. Petites will fail to grow on YPG and auxotrophs will fail to grow on SD. If all of the strains with the appropriate phenotypes are recovered, then cross-contamination is minimal.

REFERENCES

Brachmann CB, Davies A, Cost GJ, Caputo E, Li J, Hieter P, Boeke JD. 1998. Designer deletion strains derived from *Saccharomyces cerevisiae* S288C: A useful set of strains and plasmids for PCR-mediated gene disruption and other applications. *Yeast* **14:** 115–132.

Eason RG, Pourmand N, Tongprasit W, Herman ZS, Anthony K, Jejelowo O, Davis RW, Stolc V. 2004. Characterization of synthetic DNA bar codes in *Saccharomyces cerevisiae* gene-deletion strains. *Proc Natl Acad Sci* **101:** 11046–11051.

Ghaemmaghami S, Huh WK, Bower K, Howson RW, Belle A, Dephoure N, O'Shea EK, Weissman JS. 2003. Global analysis of protein expression in yeast. *Nature* **425:** 737–741.

Huh WK, Falvo JV, Gerke LC, Carroll AS, Howson RW, Weissman JS, O'Shea EK. 2003. Global analysis of protein localization in budding yeast. *Nature* **425:** 686–691.

Ooi SL, Shoemaker DD, Boeke JD. 2001. A DNA microarray-based genetic screen for nonhomologous end-joining mutants in *Saccharomyces cerevisiae*. *Science* **294:** 2552–2556.

Smith AM, Heisler LE, Mellor J, Kaper F, Thompson MJ, Chee M, Roth FP, Giaever G, Nislow C. 2009. Quantitative phenotyping via deep barcode sequencing. *Genome Res* **19:** 1836–1842.

Sopko R, Huang D, Preston N, Chua G, Papp B, Kafadar K, Snyder M, Oliver SG, Cyert M, Hughes TR, et al. 2006. Mapping pathways and phenotypes by systematic gene overexpression. *Mol Cell* **21:** 319–330.

Zhu H, Bilgin M, Bangham R, Hall D, Casamayor A, Bertone P, Lan N, Jansen R, Bidlingmaier S, Houfek T, et al. 2001. Global analysis of protein activities using proteome chips. *Science* **293:** 2101–2105.

Scoring SGA Screens with SGATools

QUANTITATIVE SCORING OF GENETIC INTERACTIONS

SGATools is a web-based analysis system that you will use to score genetic interactions (Wagih et al. 2013). It will enable you to quantify colonies on your agar plates, normalize systematic effects, and calculate fitness scores relative to a control experiment (Fig. 1). Additional analysis tools are also available such as viewing scored colonies on plates as well as the shape of the distribution of the genetic interaction scores. The website is also linked to external sources to aid with Gene Ontology (GO) enrichment analysis.

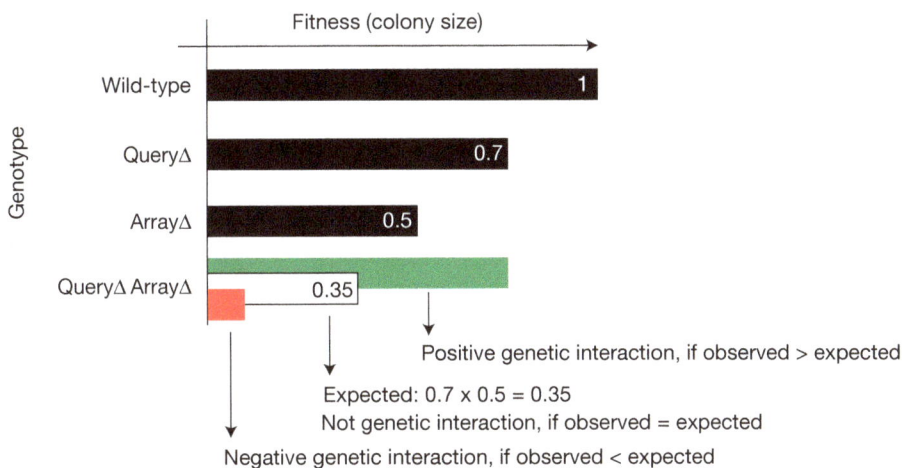

FIGURE 1. A graphical representation of how genetic interactions are measured from colony size. Wild-type fitness is defined as 1. The fitness of two single mutants are expected to combine multiplicatively. In this case, query fitness is 0.7 and array fitness is 0.5; thus, the double-mutant fitness is expected to be 0.35. If the observed colony size of the double mutant is 0.35, then we do not score it as a genetic interaction; however, if it is smaller or larger than 0.35, then we call it a negative or positive genetic interaction, respectively.

Adapted from a protocol by Elena Kuzmin and Charlie Boone, University of Toronto.

FOLLOW THESE INSTRUCTIONS TO ANALYZE YOUR SCREENS

Start by naming your images using the appropriate format. Each query mutant screen must be paired with a control screen for the analysis step. Exchange images with another group so that each pair of students has a complete set of query and control images to analyze. The query ORF names are as follows:

rad1Δ   YPL022W

rtt107Δ   YHR154W

slx1Δ   YBR228W

slx4Δ   YLR135W

The control ORF is *ura10Δ* YMR271C.

Wild-type control screen:

   YourName_wt_controlORFname_PlateNumber.jpg

Double mutant screen:

   YourName_dm_queryORFname_PlateNumber.jpg

For example, if you are screening *rad1Δ*, the images for plates 1 are named

   GrantBrown_wt_ YMR271C _001.jpg

   GrantBrown_dm_YPL022W_001.jpg

You can find the SGATools website at http://sgatools.ccbr.utoronto.ca/. You should use Google Chrome browser for the most optimal display.

To Process the Images and Obtain Colony Sizes in Pixels

1. At the top of the page click on "Image Analysis."

2. Under "Plate image(s)," click on "Select image(s)" → select both the control single-mutant array images and the double-mutant array images that you wish to analyze and click "Open." The images all need to reside in the same folder on your computer.

   a. You will need to analyze one double-mutant array at a time.

   b. Leave "Options" at default.

   c. Enter your email address.

3. Ensure that "Loaded image files" have your files and "Screen type," "Query name," and "Array plate id" identified correctly.

4. Leave the other settings at default:

   a. Under "Plate format," it should say "1536 colonies (32 × 48)."

   b. Under "Options," the "Autorotate" checkbox should be deselected and the "Colonies" pull-down should read "Bright."

   c. Enter your email address to be notified when the job is complete.

5. Click "Process images."

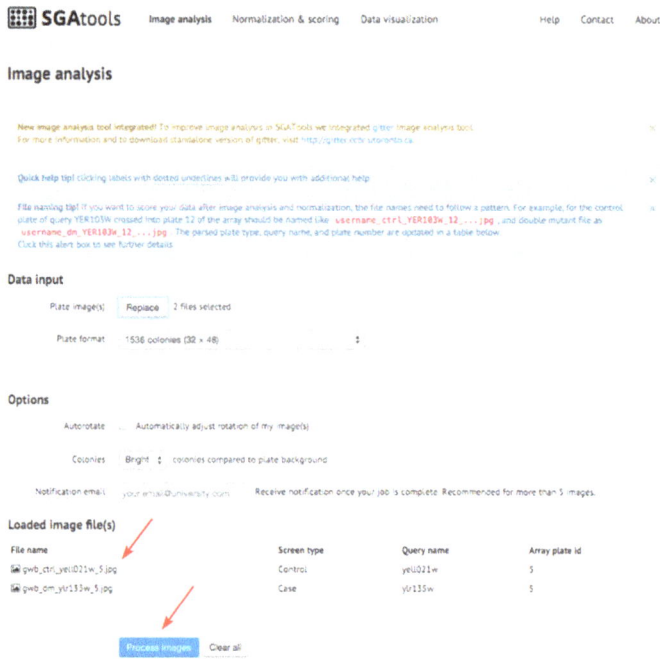

6. Verify proper segmentation of the colonies by hovering over and visually inspecting the black and white images of the arrays (especially at the edges). The "Status" of all images should say "Passed."

7. Scroll to the top of the page and click "Normalize and score."

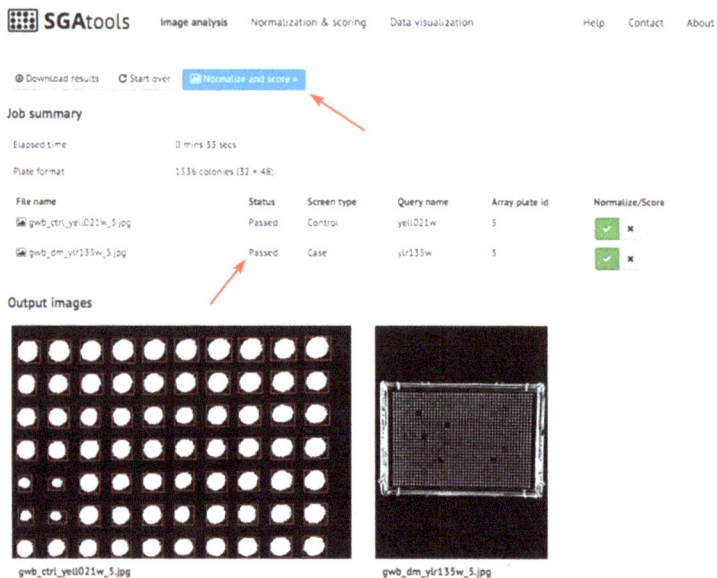

To Calculate the Normalized Colony Sizes and Score Genetic Interactions

8. Under "Data input":

   a. Ensure that the "Plate layout" checkbox is selected.

   b. Use the pull-down menu to select "sga-array-ver2-1536." The other pull-down should read "All plates."

9. Under "Options":

   a. Under "Replicates," select "4 (2 × 2)."

   b. Ensure that the "Linkage correction" checkbox is selected to "Filter out genes linked to the query." Leave "Linkage cutoff" at "200 KB" and "Linkage genes" as is.

   c. Under "Score results," ensure that "Scores the normalized output" is selected.

10. Click "Normalize and score."

11. To download results, click "Download" at the top of the page and select Excel format. You will find the genetic interaction scores in the tab "Combined data" in column "Score." Sort the spreadsheet by "Score" from lowest to highest, such that your strongest negative interactions will be at the top of the list. Remember to save the Excel file!

12. To visualize the data, click "Data analysis" under "Normalization/Scoring complete!" heading.

13. Example output. The top panel shows the image of your plate and a heatmap of the colony sizes in pixels for you to verify that image processing was correct. The bottom panel displays the distribution of colony sizes. "Value" in the table below the histogram is colony size.

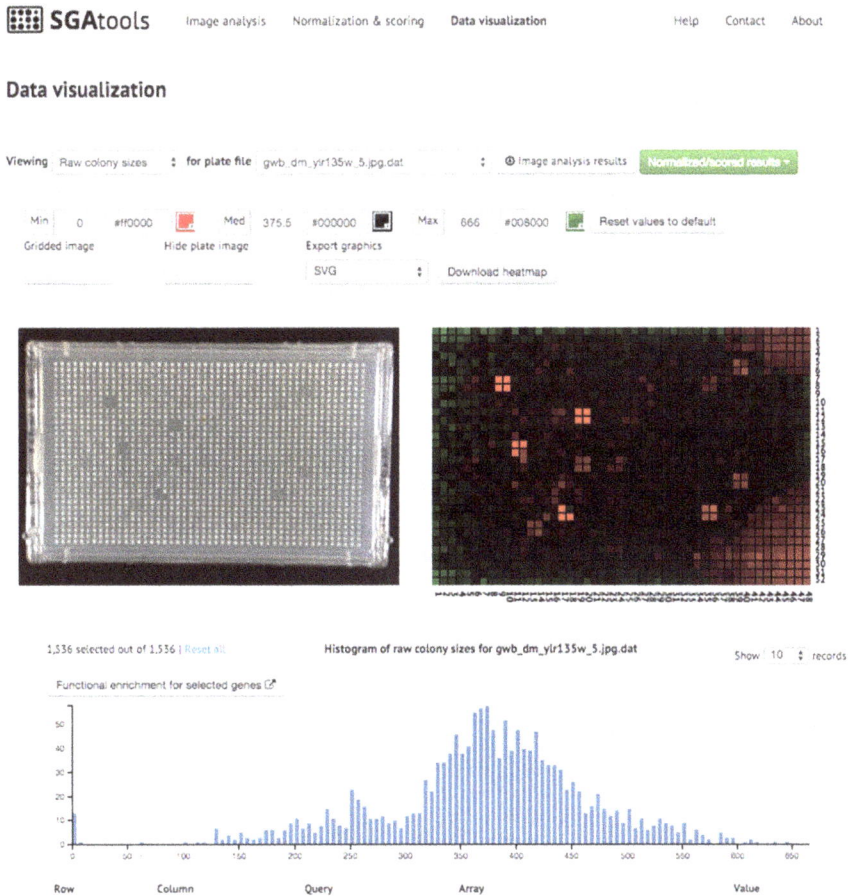

To Display Genetic Interaction Scores

14. Under "for plate file," select "Combined data."

15. Under "Viewing," select "Scored data."

16. Scroll to the distribution of genetic interaction scores and note its features.

   a. You will notice that the distribution centers at 0. This is because genetic interactions are rare and most gene pairs do not interact. The left and right tails contain negative and positive genetic interactions, respectively.

   b. "Value" in the table below the histogram represents the genetic interaction score. If a colony is sicker in the double-mutant screen than expected from the single-mutant array strains, then there is a negative genetic interaction between the query and the array strain. If a colony is healthier than expected, then there is a positive genetic interaction between the query and the array strain.

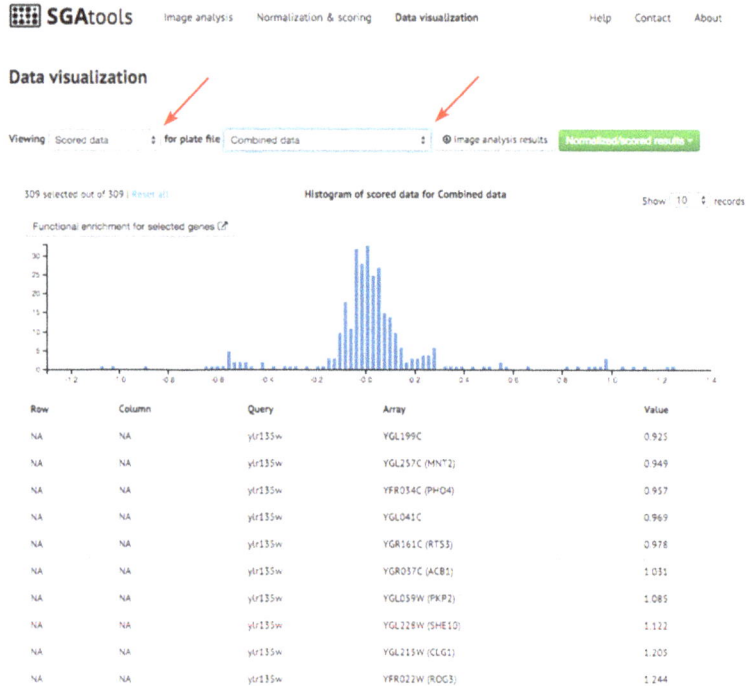

17. Select the range of genetic interaction scores, which you would like to display and analyze further, by clicking and dragging on the histogram.

    a. Scores lower than −0.3 correspond to strong negative genetic interactions that you should readily see by eye and represent instances of synthetic lethality.

    b. Scores between −0.3 and −0.1 correspond to moderate negative genetic interactions, i.e., synthetic sick effects.

    c. Scores between −0.1 and 0.1 are unlikely to represent reproducible effects.

    d. Scores between 0.1 and 0.3 should be viewed with caution because they are likely to be confounded by competition effect.

        i. Colonies around sick mutants grow larger, because of greater access to nutrients, rather than a positive genetic interaction or a fitness advantage conferred by its genotype.

    e. Scores higher than 0.3 are due to strong positive genetic interactions.

f. To see all genetic interactions at a particular cut-off, select the desired number of records to display on the right-hand side.

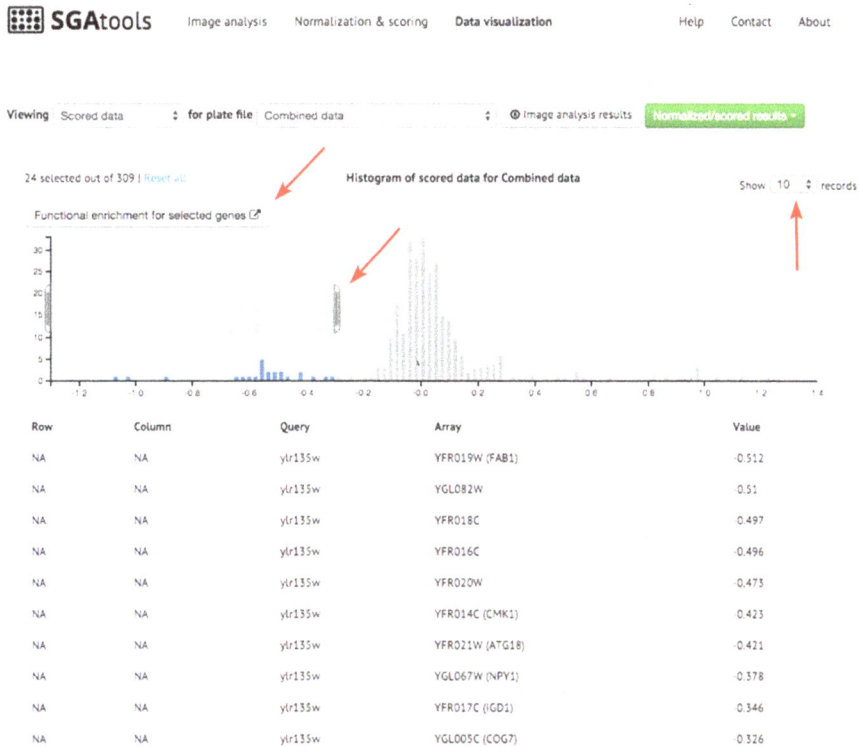

Row	Column	Query	Array	Value
NA	NA	ylr135w	YFR019W (FAB1)	-0.512
NA	NA	ylr135w	YGL082W	0.51
NA	NA	ylr135w	YFR018C	0.497
NA	NA	ylr135w	YFR016C	-0.496
NA	NA	ylr135w	YFR020W	0.473
NA	NA	ylr135w	YFR014C (CMK1)	0.423
NA	NA	ylr135w	YFR021W (ATG18)	0.421
NA	NA	ylr135w	YGL067W (NPY1)	0.378
NA	NA	ylr135w	YFR017C (IGD1)	0.346
NA	NA	ylr135w	YGL005C (COG7)	0.326

18. Think about what the genetic interactions mean based on what is already known about the query genes in the literature. To get started, use SGD.

19. GO enrichment using g:Profiler (Reimand et al. 2011).

a. Select genes of interest above a certain threshold (start from −0.3 and lower) by pressing and dragging on the histogram.

b. Click on "Functional enrichment for selected genes" on the top left of the histogram.

c. Scroll down to see which GO terms are enriched in your data as indicated by a significant p value and highlighted in black. Think about what it means with respect to the function of your query gene. To return to the output at a later date, click on ">>Static URL" and wait for the page to load. Then save it to your bookmarks. Alternatively, print out the output or save the page as a PDF.

d. To display GO terms in a hierarchical order, under "Options" at the top of the page, select "Hierarchical sorting" and then click "g:Profile!." Be patient! This will take some time to load. Scroll through the terms again to see how they are organized. You can familiarize yourself with gene ontology (GO) structure at http://www.geneontology.org/GO.ontology.structure. shtml. To return to the output at a later day, click on ">>Static URL" and wait for the page to load. Then save it to your bookmarks. Alternatively, print out the output or save the page as PDF.

REFERENCES

Reimand J, Arak T, Vilo J. 2011. g:Profiler—A web server for functional interpretation of gene lists (2011 update). *Nucleic Acids Res* **39:** W307–W315.

Wagih O, Usaj M, Baryshnikova A, VanderSluis B, Kuzmin E, Costanzo M, Myers CL, Andrews BJ, Boone CM, Parts L. 2013. SGATools: One-stop analysis and visualization of array-based genetic interaction screens. *Nucleic Acids Res* **41:** W591–W596.

Measuring DNA Concentration with the Qubit Fluorometer

TIPS

Make a little extra buffer + reagent mix every time you use the Qubit so that you do not run out. Even the standards must have the appropriate volume of master mix for accurate readings. You must recalibrate the Qubit every time you use it, with the two standards.

PROCEDURE

1. Make a master mix of the Qubit buffer + Qubit reagent (199:1) with a total of 200 µL per sample. Do not forget to make enough for the two standards!

2. Aliquot 198 µL of the master mix per test DNA sample and 190 µL of master mix per standard sample into the special 0.5-mL Qubit tubes.

3. Add 2 µL of test DNA and 10 µL of standard DNA to the appropriate tubes.

4. Vortex tubes for 3 sec.

5. Incubate tubes at room temperature (in the dark if possible) for 2 min.

6. Read the µg/mL using the Qubit machine.

7. Press the "home" button and select DNA, Broad Range (BR).

8. Select "run new calibration."

9. Follow prompts and insert standard 1, and press "Go."

10. Follow prompts and insert standard 2, and press "Go."

11. Insert test DNA sample, and press "Go" to read.

12. Using the 2 µL of DNA:198 µL of master mix dilution as described above, you must multiply the µg/mL reported by the machine by 100 to get the actual concentration of your DNA sample. Alternatively, you can type the values into the instrument using the "Calculate sample concentration" function.

Colony PCR

PROCEDURE

1. Identify a well-isolated colony.

2. Use a toothpick to transfer a dab (do not scrape out the whole colony) of yeast cells from each colony to 30 µL of 0.2% SDS.

3. Vortex the cell/SDS mixture for ~15 sec.

4. Place in heat block for 4 min at 90°C.

5. Centrifuge for 1 min and transfer the supernatant to a new tube.

6. No more than 1 µL should be added to 50 µL FastStart PCR reaction. The SDS can inhibit the reaction.

7. Set up PCR as follows:

FastStart PCR reaction buffer, 10×	5 µL
dNTP mix (10 mM each)	2 µL
FastStart *Taq* DNA polymerase (5 units/µL; Roche)	0.8 µL
Oligo1 (10 µM)	4 µL
Oligo2 (10 µM)	4 µL
DNA	1 µL
H_2O	to 50 µL

Training for the Plate Race

PROCEDURE

1. Wake up.

2. Shake off hangover from a late night of drinking with the guest lecturer.

3. Put on running shoes (brightly colored shoes work better).

4. Go back to bed.

5. Wake up again.

6. Go to lab and collect 40 plates with assorted fungi, none of them being *Saccharomyces cerevisiae*.

7. Assemble on Bungtown Road with peers and try not to look silly, as Jim Watson and Bruce Stillman drive by unfazed and disinterested.

8. Rush back to lab with the three plates that you now realize actually contain valuable information. That experiment worked after all! Go figure.

9. Return to Bungtown Road.

10. Attempt to run the loop with 40 plates in a stack.

11. Drop plates. Repeat Steps 6–11 until it is time for *Nuts and Bolts*.

12. When you have completed a full lap without a drop, you are ready to graduate from The Yeast Course.

13. Not quite yet. You must also dissect 20 tetrads successfully.

Media

M edia for Petri plates are prepared in 2-L flasks, with each flask containing 1 L of medium, which is sufficient for 30–40 plates. Unless otherwise stated, all components are autoclaved together for 15 min at 250°F (121°C) and 15 psi. Allow the plates to dry at room temperature for 2–3 d after pouring. The plates can be stored in sealed plastic bags for more than 3 mo. The agar is omitted for liquid media. (For convenience, the final concentration of each component in the medium is listed in parentheses below.)

A note about glucose: Autoclaving glucose together with the other components of media for periods longer than 15 min leads to caramelization of the sugar (noted by dark brown color) and less optimal growth of yeast. To avoid caramelization, some laboratories autoclave a stock solution of 20% glucose separately and add the sugar (100 mL/L of media) after all other ingredients have been autoclaved.

A note about mushy plates: Autoclaving minimal media longer than 15 min hydrolyzes the agar, resulting in mushy plates. If longer autoclaving times are needed, add one sodium hydroxide pellet (~0.1 g) per liter of medium. pH can be adjusted after autoclaving with 1 M HCl, although this is rarely necessary.

RICH MEDIA

YPD (YEPD)

YPD is a complex medium for routine growth

Bacto-yeast extract (1%)	10 g
Bacto-peptone (2%)	20 g
Glucose (2%)*	20 g
Bacto-agar (2%)	20 g
Distilled H$_2$O	to 1000 mL

> *Note: ade1* and *ade2* mutants accumulate a red pigment that causes a minor growth defect. YPAD media (YPD media plus 0.72 g of adenine hemisulfate) is sometimes used to suppress pigment accumulation and remove selection for secondary mutations that block pigment accumulation.

YPG (YEPG OR YEP-GLYCEROL)

YPG is a complex medium containing a nonfermentable carbon source (glycerol) that does not support the growth of ρ⁻ or *pet* mutants.

Bacto-yeast extract (1%)	10 g
Bacto-peptone (2%)	20 g
Glycerol (3% [v/v])	30 mL
Bacto-agar (2%)	20 g
Distilled H$_2$O	to 1000 mL

YEP

YEP is a complex medium to which additional carbon sources can be added. For most purposes, a final concentration of 2% sugar is sufficient. If raffinose is used, it is best to prepare a 20% stock solution by filter sterilization through a 0.2-μm Nalgene filter and add 100 mL after the autoclaved media has cooled.

Bacto-yeast extract (1%)	10 g
Bacto-peptone (2%)	20 g
Bacto-agar (2%)	20 g
Distilled H$_2$O	to 900 mL

SYNTHETIC MEDIA

Synthetic Dextrose Minimal Medium

Synthetic dextrose (SD) is a synthetic minimal medium containing salts, trace elements, vitamins, a nitrogen source (Bacto-yeast nitrogen base without amino acids), and dextrose (i.e., glucose).

Bacto-yeast nitrogen base without amino acids (0.67%)	6.7 g
Glucose (2%)	20 g
Bacto-agar (2%)	20 g
Distilled H$_2$O	to 1000 mL

Supplemented Minimal Medium (or Add-back Medium)

Supplemented minimal medium (SMM) is SD to which various nutritional supplements have been added. Typically, SMM media contains only those nutrients necessary for growth of the strain of interest. The medium can be prepared by adding nutrients to SD immediately prior to or after autoclaving. Alternatively, an appropriate volume of nutrient stock solutions can be spread on the surface of an SD plate. In this way, a variety of selective plates can be generated rapidly from a single batch of SD plates.

Stock solutions of each of the nutrients can be prepared and stored for extensive periods. Some should be stored at room temperature in order to prevent precipitation, whereas other solutions may be refrigerated. Wherever applicable, HCl salts of amino acids are preferred. After spreading nutrient solutions on the surface of plates, allow sufficient time for the liquid to absorb into the agar before adding yeast.

Table A1 contains the necessary information for creating and using stock solutions for adding nutrients back to the SD media.

Synthetic Complete and Dropout Media

Synthetic complete (SC) is SD that has been supplemented with a generic list of nutrients that are required by strains with commonly encountered auxotrophies. Most laboratory strains grow on SC, albeit more slowly than on YPD. Dropout media, on the other hand, is SC from which one or more essential nutrients has been "dropped out." Dropout media is useful when selecting for plasmids that carry biosynthetic markers following transformation. A dropout medium lacking leucine and histidine, for example, is designated SC-leu-his. A dropout medium containing all of the nutrients listed in Table A2 is designated SC.

Table A1. Nutrient supplements to add back to SD media

Constituent	Stock concentration (g/100 mL)	Volume of stock for 1 liter of medium (mL)	Final concentration in medium (mg/L)	Volume of stock to spread on plate (mL)
Adenine sulfate	0.2[a]	10	20	0.2
Uracil	0.2[a]	10	20	0.2
L-Tryptophan	1	2	20	0.1
L-Histidine HCl	1	2	20	0.1
L-Arginine HCl	1	2	20	0.1
L-Methionine	1	2	20	0.1
L-Tyrosine	0.2	15	30	0.2
L-Leucine	1	10	100	0.1
L-Isoleucine	1	3	30	0.1
L-Lysine HCl	1	3	30	0.1
L-Phenylalanine	1[a]	5	50	0.1
L-Glutamic acid	1[a]	10	100	0.2
L-Aspartic acid	1[a,b]	10	100	0.2
L-Valine	3	5	150	0.1
L-Threonine	4[a,b]	5	200	0.1
L-Serine	8	5	400	0.1

[a] Store at room temperature.
[b] Add after autoclaving medium.

Table A2. Nutrient formulations for SD media and drop-out powders

Nutrient	Amount (g)
L-Adenine	0.5
L-Alanine	2.0
L-Arginine	2.0
L-Asparagine	2.0
L-Aspartic acid	2.0
L-Cysteine	2.0
L-Glutamine	2.0
L-Glutamic acid	2.0
L-Glycine	2.0
L-Histidine	2.0
myo-Inositol	2.0
L-Isoleucine	2.0
L-Leucine	10.0
L-Lysine	2.0
L-Methionine	2.0
Para-aminobenzoic acid	2.0
L-Phenylalanine	2.0
L-Proline	2.0
L-Serine	2.0
L-Threonine	2.0
L-Tryptophan	2.0
L-Tyrosine	2.0
Uracil	2.0
L-Valine	2.0

Note: Other common formulations of nutrient mixtures contain fewer nutrients. For example, CSM (complete synthetic medium) contains only Ade, Arg, Asp, His, Ile, Leu, Lys, Met, Phe, Thr, Trp, Tyr, Ura, Val.

Dropout powders contain some or all of the ingredients listed in Table A2. The powders are prepared in bulk to avoid weighing out individual nutrients each time a stack of plates is needed. Wherever applicable, use HCl salts of amino acids. Powders should be mixed thoroughly in a blender. Alternatively, turning a vessel containing the nutrients end-over-end overnight also works; adding a couple of clean marbles helps. It is often convenient to prepare a single dropout powder lacking all nutrients required of common laboratory strains (e.g., adenine, histidine, leucine, tryptophan, uracil for strain W303) and then to add back nutrients as necessary using the stock solutions described in Supplemented Minimal Media, Table A1. Dropout powders are stable for years.

Raw materials for compounding synthetic media, or complete mixtures, can be purchased from commercial vendors (Sunrise Science [http://www.sunrisescience.com], Sigma-Aldrich [http://www.sigmaaldrich.com], and Teknova [http://www.teknova.com]).

Both SC and dropout media are assembled from yeast nitrogen base (without amino acids), a carbon source (usually glucose) and a dropout powder.

Bacto-yeast nitrogen base without amino acids (0.67%)	6.7 g
Glucose (2%)	20 g
Bacto-agar (2%)	20 g
Dropout powder (0.2%)	2 g
Distilled H_2O	to 1000 mL

SPORULATION MEDIUM

For sporulation plates, add 20 g of Bacto-agar (2%) to the recipes for liquid media provided below.

SPO

*MAT***a**/*MAT*α diploids will undergo several divisions over 3–5 d before sporulating on this medium.

Potassium acetate (1%)	10 g
Bacto-yeast extract (0.1%)	1 g
Glucose (0.05%)	0.5 g
Dropout powder (0.01%)*	0.1 g
Distilled H_2O	to 1000 mL

> *Note:* Nutritional supplements are required for sporulation of auxotrophic diploids. These can be provided by (1) adding 0.1 g of dropout powder or CSM powder or (2) adding only those nutrients required by the strain (add 25% of the level of each nutrient normally used for SMM/add-back media; Table A1).

Minimal SPO

*MAT***a**/*MAT*α diploid cells will sporulate on this medium after 18–24 h without vegetative growth. It is referred to as minimal SPO because it lacks glucose and yeast extract.

Potassium acetate (1%)	10 g
Dropout powder (0.01%)*	0.1 g
Distilled H_2O	to 1000 mL

> *Note:* Nutritional supplements are required for sporulation of auxotrophic diploids. These can be provided by (1) adding 0.1 g of dropout powder or CSM powder or (2) adding only those nutrients required by the strain (add 25% of the level of each nutrient normally used for SMM/add-back media; Table A1).

Presporulation Media

Some strains (e.g., BY474) sporulate more efficiently if pregrown in rich YPD (YEP + 4% glucose) prior to transfer to SPO media.

Yeast extract (1%)	10 g
Bacto-peptone (2%)	20 g
Glucose (4%)	40 g
Distilled H$_2$O	to 1000 mL

> *Note:* SK1 cells are typically first grown in YPEG (YEP + 3% glycerol + 2% ethanol) before growth in YPA (YEP + 2% acetate) for optimal sporulation. However, SK1 will sporulate adequately even if pregrown in YPD.

INDICATOR MEDIA

Limiting Ade Plates

The *ADE2* gene encodes phosphoribosylamino-imidazole-carboxylase, which is essential for the synthesis of adenine. Mutations in *ade2* result in the accumulation of the red-pigmented 5-phosphoribosyl-5-aminoimidazole. Cells containing a wild-type copy of *ADE2* are cream colored because this compound is quickly metabolized. Mutations in genes that function upstream of Ade2, such as *ade3*Δ, will also be cream colored because synthesis of 5-phosphoribosyl-5-aminoimidazole is blocked at an earlier step.

To observe the color changes associated with Ade2, cells are grown on plates containing limiting amounts of adenine. For certain applications, the low level of adenine in YPD plates is sufficient for color detection. After the colonies appear, a short time in the refrigerator will help the red pigment develop. When a precise adenine level is desired or a synthetic dropout media is required, assemble the following media with limiting adenine:

Bacto-yeast nitrogen base without amino acids (0.67%)	6.7 g
Glucose (2%)	20 g
Bacto-agar (2%)	20 g
SC-Ade dropout powder (0.2%)	2 g
Distilled H$_2$O	to 1000 mL

After autoclaving, cool media to 55°C and add adenine to 5 mg/L* Mix and pour.

> *Note:* The amount of adenine required to achieve optimal red color may vary from one strain to the next.

MAL Indicator Plates

MAL indicator medium is a fermentation-indicator medium used to distinguish strains that ferment maltose from those that do not. Due to the pH change, the maltose-fermenting strains will change the indicator yellow.

Bacto-yeast extract (1%)	10 g
Bacto-peptone (2%)	20 g
Maltose (2%)	20 g
Bromcresol purple solution (0.036%)	9 mL of 4% stock
Bacto-agar (2%)	20 g
Distilled H_2O	to 1000 mL

0.4% Bromcresol purple solution:	
Bromcresol purple	200 mg
100% Ethanol	50 mL

X-Gal Plates for Lysed Yeast Cells on Filters

These plates are used for checking β-galactosidase activity in cells that have been lysed and are immobilized on Whatman 3MM filters.

Bacto-agar	20 g
1 M Na_2HPO_4	57.7 mL
1 M NaH_2PO_4	42.3 mL
$MgSO_4$	0.25 g
Distilled H_2O	to 1000 mL

After autoclaving, add 6 mL of X-Gal solution (20 mg/mL in *N,N*-dimethylformamide).

DRUG SELECTION MEDIA

5-Fluoro-Orotic Acid Plates

5-Fluoro-orotic acid (5-FOA) is used in a variety of assays as a counterselective agent for *URA3*. In wild-type cells, the gene products of *URA5* (orotate phosphoribosyltransferase) and *URA3* (orotidine-5′-phosphate decarboxylase) sequentially convert 5-FOA to 5-fluoro-uridine monophosphate (5-FUMP). Further biosynthetic modification yields fluorodeoxyuridine, which is a potent and toxic inhibitor of thymidylate synthase. *ura3* and *ura5* mutants can grow on medium containing 5-FOA as long as uracil is provided to form UMP via the pyrimidine salvage pathway (Boeke et al. 1984).

Dissolve the following reagents in H_2O and filter-sterilize using a 0.2-μm filter:

Bacto-yeast nitrogen base without amino acids (0.67%)	6.7 g
Glucose (2%)	20 g
SC-ura dropout powder (0.2%)	2 g
Uracil (50 µg/L)*	50 mg
5-FOA (0.1%)	1 g
Distilled H$_2$O	to 500 mL

Note: Uracil can be added from a sterile stock solution after autoclaving.

Autoclave the following reagents separately:

Bacto-agar (2%)	20 g
Distilled H$_2$O	500 mL

Mix the two solutions after cooling the agar to ~80°C. Pour into Petri dishes.

α-Aminoadipate Plates

Wild-type strains are unable to use high levels of α-aminoadipate (α-AA) as their sole nitrogen source because it is converted into a toxic intermediate by the gene products of *LYS2* and *LYS5* in the normal lysine anabolic pathway (Chattoo et al. 1979; Zaret and Sherman 1985). Because cells that grow on α-AA will be lysine auxotrophs, lysine must be added to the culture medium. α-AA is frequently used for counter-selection of *LYS2*.

Autoclave the following separately:

Bacto-YNB without amino acids and without ammonium sulfate (0.17%)	1.7 g
Glucose (2%)	20 g
Bacto-agar (2%)	20 g
Lysine (30 mg/L)*	30 mg
Distilled H$_2$O	to 960 mL

Autoclave and add 40 mL of a 5% α-AA solution. Add back those nutritional supplements necessary for growth if your strain is auxotrophic.

Note: Lysine can be added from a sterile stock solution after autoclaving.

5% α-AA:

α-Aminoadipic acid	2 g
Distilled H$_2$O	to 40 mL

Mix and adjust the pH to 6 with 10 N KOH to allow dissolution. Filter-sterilize using a 0.2-μm filter before adding to the autoclaved ingredients.

5-Fluoroanthranilic Acid (5-FAA) Medium

5-Fluoroanthranilic acid (2-amino-5-fluorobenzoic acid) is used for counter selection of *TRP1*, which encodes phosphoribosylanthranilate isomerase of the tryptophan biosynthetic pathway. Wild-type cells convert 5-FAA to 5-fluorotryptophan, which is toxic to yeast. Because cells that grow on 5-FAA will be tryptophan auxotrophs, tryptophan must be added to the culture media. Selection on 5-FAA is enhanced by increased glucose and amino acid concentrations (Toyn et al. 2000). Autoclave the following separately:

Bacto-yeast nitrogen base without amino acids (0.67%)	6.7 g
Glucose (5%)	50 g
Bacto-agar (2%)	20 g
SC-trp dropout powder (0.2%)	2 g
Tryptophan (10 mg/L)*	10 mg
Distilled H$_2$O	to 1000 mL

After autoclaving, cool to 55°C and add 5 mL of 10% 5-FAA. Stir and then pour.

Note: Tryptophan can be added from a sterile stock solution after autoclaving.

10% 5-FAA:

5-FAA	1 g
Ethanol	10 mL

Filter sterilize and store in 5-mL aliquots at −20°C.

Cycloheximide

Cycloheximide resistance can arise in a number of different genes. However, resistance to high levels ordinarily arises from rare mutations at the *cyh2* locus, which encodes the L29 ribosomal subunit. Resistance to cycloheximide is recessive, presumably because the drug blocks further elongation and traps sensitive ribosomes on mRNA.

Cycloheximide can be used in either YPD or synthetic media. A final concentration of 10 mg/L should be used for YPD and 3 mg/L for SD, SC, and YPG. A stock solution is prepared by dissolving 100 mg of cycloheximide in 10 mL of distilled H$_2$O and then filter-sterilizing (0.2-μm filter). The stock solution can be stored at 4°C. Appropriate volumes can be added to the media after autoclaving.

Canavanine

Canavanine is an analog of arginine. Both are imported into the cell via the same high-affinity permease encoded by the *CAN1* locus. Mutation of this locus confers high-level resistance to canavanine, whereas mutation of a number of other loci confers low-level resistance.

Because arginine competes with canavanine for Can1, arginine must be excluded from media used for testing drug sensitivity. Canavanine resistance must also be scored in high-nitrogen media (e.g., SD or SC) in which Can1 is the only route for arginine and canavanine to enter the cell. Low-nitrogen conditions (e.g., YPD) induce the general amino acid permease (Gap1), which can also take up arginine and canavanine. For this reason, Can^R Arg^- auxotrophs are viable on YPD but not synthetic media, even when it contains arginine.

Canavanine-L-sulfate is typically made up as a filter-sterilized (0.2-μm filter) 100 mg/mL stock solution in distilled H_2O. It is stored at 4°C and added to SD or SC–arg medium after autoclaving. A concentration of 50 mg/L is typically used for scoring and selecting canavanine resistance.

Thialysine

Thialysine (*S*-(2-aminoethyl)-L-cysteine hydrochloride) is an analog of lysine, and it is imported into the cell via the Lyp1 permease. Much like canavanine and arginine, lysine and thialysine compete for Lyp1, and *LYP1* expression is regulated by the availability of lysine. Therefore, lysine must be removed from the media to score thialysine resistance.

Thialysine hydrochloride is typically made up as a filter-sterilized (0.2-μm filter) 100 mg/mL stock solution in distilled H_2O. It is stored at 4°C and added to SD or SC–lys medium after autoclaving. A concentration of 50 mg/L is typically used for scoring and selecting thialysine resistance.

G418

The *kan* gene from the *Escherichia coli* transposon Tn*903* confers resistance to G418. This gene has been engineered into *KANMX* modules that can be used for polymerase chain reaction (PCR)-based gene disruption and tagging. G418 disulfate salt is made as a filter-sterilized (0.2-μm filter) 200 mg/mL stock solution in distilled H_2O. It is stored in aliquots at 4°C. To select for *KANMX* transformants, YPD is supplemented with 200 mg/L G418. Note that kanamycin suitable for bacterial selection is not effective in yeast.

G418 is not effective in minimal media containing ammonium sulfate. When minimal media is needed, replace the ammonium sulfate with 1 g/L monosodium glutamate. SC/MSG + G418:

Bacto-YNB without amino acids and without ammonium sulfate (0.17%)	1.7 g
Glucose (2%)	20 g
Bacto-agar (2%)	20 g
Dropout powder (0.2%)	2 g
Monosodium glutamate (0.1%)	1 g
Distilled H_2O	to 1000 mL

Autoclave and cool to 55°C. Add 1 mL of G418 stock solution, stir, and then pour.

clonNAT

The *Streptomyces noursei nat1* gene encodes nourseothricin *N*-acetyltransferase, which confers resistance to nourseothricin. This gene has been engineered into *NATMX* modules that can be used for PCR-based gene disruption and tagging. The commercially available version of nourseothricin is clonNAT (Werner Bio-Agents, www.webioage.de/eng/index.php). It is made as a filter-sterilized (0.2-µm filter) 100 mg/mL stock solution in distilled H_2O. It is stored in aliquots at 4°C. To select for *NATMX* transformants, YPD is supplemented with 100 mg/L clonNAT.

clonNAT is not effective in minimal media containing ammonium sulfate. When minimal media is required, replace the ammonium sulfate with 1 g/L monosodium glutamate.

Hygromycin

The *Klebsiella pneumoniae hph* gene confers resistance to hygromycin B. This gene has been engineered into *HPHMX* modules that can be used for PCR-based gene disruption and tagging. Hygromycin B is made as a filter-sterilized (0.2-µm filter) 300 mg/mL stock solution in distilled H_2O. It is stored in aliquots at 4°C. To select for *HPHMX* transformants, YPD is supplemented with 300 mg/L hygromycin B.

Hygromycin is not effective in minimal media containing ammonium sulfate. When minimal media is required, replace the ammonium sulfate with 1 g/L monosodium glutamate as described above for G418.

Phleomycin/Zeocin

The *ble* gene from the *K. pneumoniae* Tn*5* transposon confers resistance to phleomycin or zeocin. This gene has been engineered into *BLEMX* modules that can be used for PCR-based gene disruption and tagging. To select for *BLEMX* transformants, YPD is supplemented with 7.5 mg/L phleomycin or 100 mg/L zeocin. Follow manufacturer's recommendations.

In *Schizosaccharomyces pombe*, phleomycin is not effective in minimal media containing ammonium sulfate (Benko and Zhao 2011). Thus, we recommend replacing the ammonium sulfate with 1 g/L monosodium glutamate when phleomycin is required.

Bialaphos

Bialaphos is a peptide composed of two alanine residues and the glutamine analogue, phosphinothricin, which inhibits glutamine synthase. In order for Bialaphos to be taken up into the cell, the media must not contain peptides and amino acids that would normally compete for Bialaphos for transport. The *pat* gene from *Streptomyces viridochromogenes* encodes phosphinothricin *N*-acetyltransferase, which confers resistance to Bialaphos. This gene has been engineered into *PATMX* modules that can be used for PCR-based gene disruption and tagging. Bialaphos can be purchased from Toku-E. To select for *PATMX* transformants, the following SDP media is used:

Bacto-YNB without amino acids and without ammonium sulfate (0.17%)	1.7 g
Glucose (2%)	20 g
Bacto-agar (2%)	20 g
Proline (0.1%)	10 g
Distilled H_2O	to 1000 mL

After autoclaving, add Bialaphos to 200 mg/L. Pour. Add back those nutritional supplements necessary for growth if your strain is auxotrophic. Note that glufosinate can be used in place of Bialaphos at a concentration of 600–800 mg/L.

Fluoroacetamide/Acetamide

Budding yeast can be engineered to use acetamide as a sole source of both nitrogen and carbon by transformation with the *amdS* gene of *Aspergillus nidulans*. The gene encodes acetamidase, which hydrolyzes acetamide to acetate and ammonia. *amdS* has been engineered into a module (*amdSYM*) that can be used for PCR-based gene disruption (Solis-Escalante et al. 2013). *amdSYM* can be selected for positively by growth on acetamide or negatively by growth on fluoroacetamide, which is converted to the toxic compound fluoroacetate.

Acetamide Media

Bacto-YNB without amino acids and without ammonium sulfate (0.17%)	1.7 g
Glucose (2%)	20 g
Bacto-agar (2%)	20 g
Distilled H_2O	to 900 mL

After autoclaving, add a filter-sterilized solution of acetamide (0.6 g/100 mL). Add back those nutritional supplements necessary for growth if your strain is auxotrophic. Stir and then pour.

Fluoroacetamide Media

Bacto-yeast nitrogen base without amino acids (0.67%)	6.7 g
Glucose (2%)	20 g
Bacto-agar (2%)	20 g
Distilled H_2O	to 900 mL

After autoclaving, add a filter-sterilized solution of fluoroacetamide (2.3 g/100 mL). Add back those nutritional supplements necessary for growth if your strain is auxotrophic. Stir and then pour.

SGA MEDIA

All the defined SGA media are more like SC than SD. The dropout mixes are based on the SC amino acid mix (NOT on CSM or other formulations), and this formulation is important for uniform colony growth on high-density arrays (Tong and Boone 2007).

SPO + YE + Glucose + Amino Acids (1 L)

Potassium acetate	10 g
Yeast extract	1 g
Glucose	0.5 g
Amino acid mix (2 g Ura, 2 g His, 2 g Lys, 10 g Leu)	0.1 g
Agar	20 g

Dissolve in 1 L of H_2O and autoclave. Cool to 55°C and add 250 µL of 200 mg/mL G418. Stir and pour.

SD/MSG -HIS -ARG -LYS + Canavanine, Thialysine (1 L)

YNB without amino acids and ammonium sulfate	1.7 g
Monosodium glutamic acid	1 g
Amino acids mix -HIS -ARG -LYS (see Table A3)	2 g

Dissolve in 100 mL of H_2O; filter-sterilize

Agar   20 g

Dissolve in 850 mL of H_2O and autoclave. After autoclaving, add the YNB/MSG/ Amino acids mix with the agar. Add 50 mL of 40% glucose. Cool to 55°C and add 0.5 mL of 100 mg/mL canavanine and 0.5 mL of 100 mg/mL thialysine. Stir and pour.

SD/MSG -HIS -ARG -LYS + Canavanine, Thialysine, G418 (1 L)

YNB without amino acids and ammonium sulfate	1.7 g
Monosodium glutamic acid	1 g
Amino acids mix -HIS -ARG -LYS (see Table A3)	2 g

Dissolve in 100 mL of H_2O; filter-sterilize.

Agar   20 g

Dissolve in 850 mL of H_2O and autoclave. After autoclaving, add the YNB/MSG/ Amino acids mix with the agar. Add 50 mL of 40% glucose. Cool to 55°C and add 0.5 mL of 100 mg/mL canavanine, 0.5 mL of 100 mg/mL thialysine, 1 mL of 200 mg/mL G418. Stir and pour.

SD/MSG -HIS -ARG -LYS + Canavanine, Thialysine, G418, NAT (1L)

YNB without amino acids and ammonium sulfate	1.7 g
Monosodium glutamic acid	1 g
Amino acids mix -HIS -ARG -LYS (see Table A3)	2 g

Dissolve in 100 mL of H_2O; filter-sterilize.

Agar   20 g

Dissolve in 850 mL of H_2O and autoclave. After autoclaving, add the YNB/MSG/ Amino acids mix with the agar. Add 50 mL of 40% glucose. Cool to 55°C and add 0.5 mL of 100 mg/mL canavanine, 0.5 mL of 100 mg/mL thialysine, 1 mL of 200 mg/mL G418, and 1 mL of 100 mg/mL clonNAT. Stir and pour.

G418 (Invitrogen 11811–031): Dissolve in H_2O at 200 mg/mL, filter-sterilize, and store in 1 mL aliquots at 4°C.

ClonNAT (Werner Bioagents 5002000): Dissolve in H_2O at 100 mg/mL, filter-sterilize, and store in 1 mL aliquots at 4°C.

Canavanine-L-sulfate (Sigma C9758–5G): Dissolve in H_2O at 100 mg/mL, filter-sterilize, and store in 1 mL aliquots at 4°C.

Thialysine (S-(2-aminoethyl)-L-cysteine hydrochloride) (Sigma A2636–5G): Dissolve in H_2O at 100 mg/mL, filter-sterilize, and store in 1 mL aliquots at 4°C.

Table A3. Amino acids mix for SGA (the amino acids to be omitted are indicated for each media type)

Nutrient	Amount (g)
Adenine	3
Alanine	2
Arginine	2
Asparagine	2
Aspartic acid	2
Cysteine	2
Glutamine	2
Glutamic acid	2
Glycine	2
Histidine	2
Inositol	2
Isoleucine	2
Leucine	10
Lysine	2
Methionine	2
Para-aminobenzoic acid	0.2
Phenylalanine	2
Proline	2
Serine	2
Threonine	2
Tryptophan	2
Tyrosine	2
Uracil	2
Valine	2

REFERENCES

Benko Z, Zhao RY. 2011. Zeocin for selection of bleMX6 resistance in fission yeast. *Biotechniques* **51:** 57–60.

Boeke JD, LaCroute F, Fink GR. 1984. A positive selection for mutants lacking orotidine-5′-phosphate decarboxylase activity in yeast: 5-Fluoro-orotic acid resistance. *Mol Gen Genet* **197:** 345–346.

Chattoo BB, Sherman F, Azubalis DA, Fjellstedt TA, Mehnert D, Ogur M. 1979. Selection of lys2 mutants of the yeast *Saccharomyces cerevisiae* by the utilization of α-aminoadipate. *Genetics* **93:** 51–65.

Tong A, Boone C. 2007. High-throughput strain construction and systematic synthetic lethal screening in *Saccharomyces cerevisiae*. *Methods Microbiol* **36:** 369–386.

Toyn JH, Gunyuzlu PL, White WH, Thompson LA, Hollis GF. 2000. A counterselection for the tryptophan pathway in yeast: 5-Fluoroanthranilic acid resistance. *Yeast* **16:** 553–560.

Solis-Escalante D, Kuijpers NG, Bongaerts N, Bolat I, Bosman L, Pronk JT, Daran JM, Daran-Lapujade P. 2013. amdSYM, a new dominant recyclable marker cassette for *Saccharomyces cerevisiae*. *FEMS Yeast Res* **13:** 126–139.

Zaret KS, Sherman F. 1985. α-Aminoadipate as a primary nitrogen source for *Saccharomyces cerevisiae* mutants. *J Bacteriol* **162:** 579–583.

Tetrad Dissection Sheets

This sheet is useful for noting the location of each spore from each tetrad, as indicated on the index grid on the SporePlay. Each row can accommodate two tetrads, with a spore deposited at positions A through D (tetrad 1) and at positions E through H (tetrad 2). Every time you put down a spore, put a "+" in the sheet. If you have a position where two or more spores would not dissect apart, put a "−" in the sheet. If the position has no spore, leave the box blank.

Plate ID#:

*MAT*a **strain:**

*MAT*α **strain:**

Date:

	A	B	C	D	E	F	G	H
Row 1								
Row 2								
Row 3								
Row 4								
Row 5								
Row 6								
Row 7								
Row 8								
Row 9								
Row 10								

This sheet is useful for recording the phenotypes observed for each tetrad. It also has a section to record the ditype for each pair of markers in the cross.

Parent:	MAT	37C	ade	ura	trp	mCherry	mTurquoise
Strain 4–6	a	−	−	+	+	+	−
Strain 4–7	alpha	+	+	+	−	−	+

Ditype columns: 37 - ade, trp - 37, ura, MAT - trp, MAT - 37, trp - ade (each recording PD / NPD / TT)

Phenotype columns: MAT, 37C, ade, ura, trp, mCherry, mTurquoise

Tetrad		
1	A B C D	
2	A B C D	
3	A B C D	
4	A B C D	
5	A B C D	
6	A B C D	
7	A B C D	
8	A B C D	
9	A B C D	
10	A B C D	

96-Well Plate Template

Plate ID:_____ Date:_____

Person:_____ Strain:_____

	1	2	3	4	5	6	7	8	9	10	11	12	
A													A
B													B
C													C
D													D
E													E
F													F
G													G
H													H
	1	2	3	4	5	6	7	8	9	10	11	12	

Templates for Making Streaks and Patches

Tetrad analysis

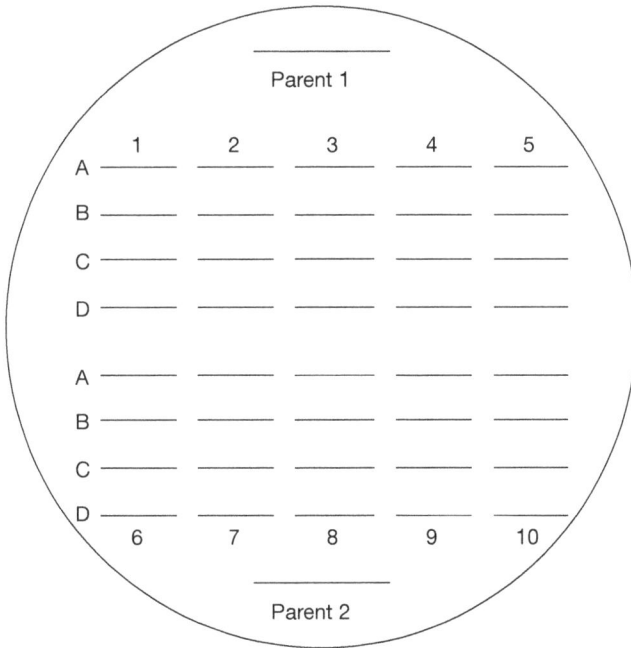

Parent 1

	1	2	3	4	5
A					
B					
C					
D					
A					
B					
C					
D	6	7	8	9	10

Parent 2

Patching

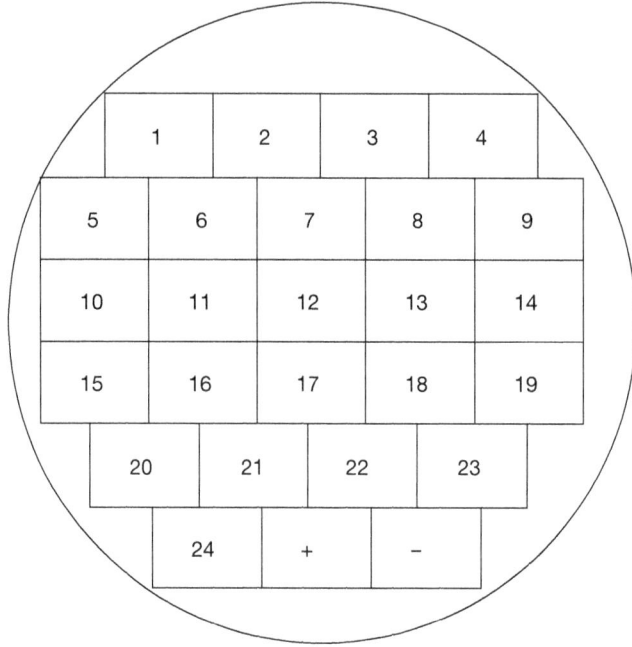

Complementation analysis

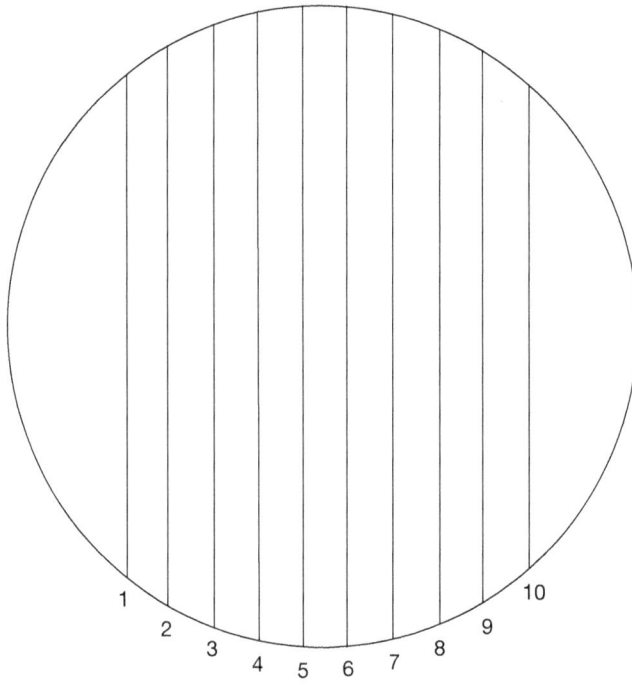

General Safety and Hazardous Material Information

> This manual should be used by laboratory personnel with experience in laboratory and chemical safety or students under the supervision of such trained personnel. The procedures, chemicals, and equipment referenced in this manual are hazardous and can cause serious injury unless performed, handled, and used with care and in a manner consistent with safe laboratory practices. Students and researchers using the procedures in this manual do so at their own risk. It is essential for your safety that you consult the appropriate Material Safety Data Sheets, the manufacturers' manuals accompanying equipment, and your institution's Environmental Health and Safety Office, as well as the General Safety and Disposal Cautions in this appendix for proper handling of hazardous materials in this manual. Cold Spring Harbor Laboratory makes no representations or warranties with respect to the material set forth in this manual and has no liability in connection with the use of these materials.
>
> All registered trademarks, trade names, and brand names mentioned in this book are the property of the respective owners. Readers should please consult individual manufacturers and other resources for current and specific product information.
>
> Appropriate sources for obtaining safety information and general guidelines for laboratory safety are provided in the General Safety and Hazardous Material Information Appendix below.

Users should always consult individual manufacturers, the manufacturers' safety guidelines, and other resources, including local safety offices, for current and specific product information and for guidance regarding the use and disposal of hazardous materials.

PRIMARY SAFETY INFORMATION RESOURCES FOR LABORATORY PERSONNEL

Institutional Safety Office

The best source of toxicity, hazard, storage, and disposal information is your institutional safety office, which maintains and makes available the most current information. Always consult this office for proper use and disposal procedures.

Post the phone numbers for your local safety office, security office, poison control center, and laboratory emergency personnel in an obvious place in your laboratory.

Material Safety Data Sheets (MSDSs)

The Occupational Safety and Health Administration (OSHA) requires that Material Safety Data Sheets (MSDSs) accompany all hazardous products that are shipped. These data sheets contain detailed safety information. MSDSs should be filed in the laboratory in a central location as a reference guide.

GENERAL SAFETY AND DISPOSAL CAUTIONS

The guidance offered here is intended to be generally applicable. However, proper waste disposal procedures vary among institutions; therefore, always consult your local safety office for specific instructions. All chemically constituted waste must be disposed of in a suitable container clearly labeled with the type of material it contains and the date the waste was initiated.

It is essential for laboratory workers to be familiar with the potential hazards of materials used in laboratory experiments and to follow recommended procedures for their use, handling, storage, and disposal.

The following general cautions should always be observed.

- **Before beginning the procedure**, become completely familiar with the properties of substances to be used.

- **The absence of a warning** does not necessarily mean that the material is safe, because information may not always be complete or available.

- **If exposed to toxic substances**, contact your local safety office immediately for instructions.

- **Use proper disposal procedures** for all chemical, biological, and radioactive waste.

- **For specific guidelines on appropriate gloves to use**, consult your local safety office.

- **Handle concentrated acids and bases** with great care. Wear goggles and appropriate gloves. A face shield should be worn when handling large quantities.

  Do not mix strong acids with organic solvents because they may react. Sulfuric acid and nitric acid especially may react highly exothermically and cause fires and explosions.

  Do not mix strong bases with halogenated solvents because they may form reactive carbenes that can lead to explosions.

- **Handle and store pressurized gas containers** with caution because they may contain flammable, toxic, or corrosive gases; asphyxiants; or oxidizers. For proper procedures, consult the Material Safety Data Sheet that is required to be provided by your vendor.

- **Never pipette** solutions using mouth suction. This method is not sterile and can be dangerous. Always use a pipette aid or bulb.

- **Keep halogenated and nonhalogenated** solvents separately (e.g., mixing chloroform and acetone can cause unexpected reactions in the presence of bases). Halogenated solvents are organic solvents such as chloroform, dichloromethane, trichlorotrifluoroethane, and dichloroethane. Nonhalogenated solvents include pentane, heptane, ethanol, methanol, benzene, toluene, *N,N*-dimethylformamide (DMF), dimethylsulfoxide (DMSO), and acetonitrile.

- **Laser radiation**, visible or invisible, can cause severe damage to the eyes and skin. Take proper precautions to prevent exposure to direct and reflected beams. Always follow the manufacturer's safety guidelines and consult your local safety office. See caution below for more detailed information.

- **Flash lamps**, because of their light intensity, can be harmful to the eyes. They also may explode on occasion. Wear appropriate eye protection and follow the manufacturer's guidelines.

- **Photographic fixatives, developers, and photoresists** also contain chemicals that can be harmful. Handle them with care and follow the manufacturer's directions.

- **Power supplies and electrophoresis equipment** pose serious fire hazard and electrical shock hazards if not used properly.

- **Microwave ovens and autoclaves** in the laboratory require certain precautions. Accidents have occurred involving their use (e.g., when melting agar or Bacto Agar stored in bottles or when sterilizing). If the screw top is not completely removed and there is inadequate space for the steam to vent, the bottles can explode and cause severe injury when the containers are removed from the microwave or autoclave. Always completely remove bottle caps before microwaving or autoclaving. An alternative method for routine agarose gels that do not require sterile agar is to weigh out the agar and place the solution in a flask.

- **Ultrasonicators** use high-frequency sound waves (16–100 kHz) for cell disruption and other purposes. This "ultrasound," conducted through air, does not pose a direct hazard to humans, but the associated high volumes of audible sound can cause a variety of effects, including headache, nausea, and tinnitus. Direct contact of the body with high-intensity ultrasound (not medical imaging equipment) should be avoided. Use appropriate ear protection and display signs on the door(s) of laboratories where the units are used.

- **Use extreme caution when handling cutting devices,** such as microtome blades, scalpels, razor blades, or needles. Microtome blades are extremely sharp! Use care when sectioning. If unfamiliar with their use, have an experienced user demonstrate proper procedures. For proper disposal, use the "sharps" disposal container in your laboratory. Discard used needles *unshielded*, with the syringe still attached. This prevents injuries and possible infections when manipulating used needles because many accidents occur while trying to replace the needle shield. Injuries may also be caused by broken pasteur pipettes, coverslips, or slides.

- **Procedures for the humane treatment of animals** must be observed at all times. Consult your local animal facility for guidelines. Animals, such as rats, are known to induce allergies that can increase in intensity with repeated exposure. Always wear a lab coat and gloves when handling these animals. If allergies to dander or saliva are known, wear a mask.

DISPOSAL OF LABORATORY WASTE

There are specific regulatory requirements for the disposal of all medical waste and biological samples mandated by the U.S. Environmental Protection Agency (see http://www.epa.gov/epawaste/hazard/tsd/index.htm) and regulated by the individual states and territories (see http://www.epa.gov/epawaste/wyl/stateprograms.htm). Medical and biological samples that require special handling and disposal are generally termed medical pathological waste (MPW), and medical, veterinary, and biological facilities will have programs for the collection of MPW and its disposal. Restrictions on how radioactive waste can be disposed of as regulated by the U.S. Nuclear Regulatory Commission can be found in 10 CFR 20.2001, General requirements for waste disposal (see http://www.nrc.gov/reading-rm/doc-collections/cfr/part020/part020-2001.html) or the individual **Agreement States**. The preferred method for the disposal of radioactively contaminated MPW is decay-in-storage (see http://www.nrc.gov/reading-rm/doc-collections/cfr/part035/part035-0092.html).

Waste and any materials contaminated with biohazardous materials must be decontaminated and disposed of as regulated medical waste. No harmful substances should be released into the environment in an uncontrolled manner. This includes all tissue samples, needles, syringes, scalpels, etc. Be sure to contact your institution's safety office concerning the proper practices associated with the handling and disposal of biohazardous waste.

Some basic rules are outlined below. For treatment of radioactive and biological waste, see sections on Radioactive Safety Procedures and Biological Safety Procedures.

- In practice, only **neutral aqueous solutions** without heavy metal ions and without organic solvents can be poured down the drain (e.g., most buffers). Acid and basic aqueous solutions need to be neutralized cautiously before their disposal by this method.

- For proper disposal of **strong acids and bases**, dilute them by placing the acid or base onto ice and neutralize them. Do not pour water into them. If the solution does not contain any other toxic compound, the salts can be flushed down the drain.

- For disposal of **other liquid waste**, similar chemicals can be collected and disposed of together, whereas chemically different wastes should be collected separately. This avoids chemical reactions between components of the mixture (see above). Collect at least inorganic aqueous waste, nonhalogenated solvents, and halogenated solvents separately.

- Waste **from photo processing and automatic developers** should be collected separately to recycle the silver traces found in it.

RADIOACTIVE SAFETY PROCEDURES

In the United States and other countries, the access to radioactive substances is strictly controlled. You may be required to become a registered user (e.g., by attending a mandatory seminar and receiving a personal dosimeter). A convenient calculator to perform routine radioactivity calculations can also be found at http://www.graphpad.com/quickcalcs/ChemMenu.cfm.

If you have never worked with radioactivity before, follow the steps below.

- **Try to avoid it!** Many experiments that are traditionally performed with the help of radioactivity can now be done using alternatives based on fluorescence or chemiluminescence and colorimetric assays, including, for example, DNA sequencing, Southern and northern blots, and protein kinase assays. However, in other cases (e.g., metabolic labeling of cells), use of radioactivity cannot be avoided.

- **Be informed.** While planning an experiment that involves the use of radioactivity, include the physicochemical properties of the isotope (half-life, emission type, and energy), the chemical form of the radioactivity, its radioactive concentration (specific activity), total amount, and its chemical concentration. Order and use only as much as is really needed.

- **Familiarize yourself** with the designated working area. Perform a mental and practical dry run (replacing radioactivity with a colored solution) to make sure that all equipment needed is available and to get used to working behind a shield. Handle your samples as if sterility would be required to avoid contamination.

- **Always wear appropriate gloves**, lab coat, and safety goggles when handling radioactive material.

- **Check the work area** for contamination before, during, and after your experiment (including your lab coat, hands, and shoes).

- **Localize your radioactivity.** Avoid formation of aerosols or contamination of large volumes of buffers.

- **Liquid scintillation cocktails** are often used to quantitate radioactivity. They contain organic solvents and small amounts of organic compounds. Try to avoid contact with the skin. After use, they should be regarded as radioactive waste; the filled vials are usually collected in designated containers, separate from other (aqueous) liquid radioactive waste.

- **Dispose of radioactive waste** only into designated, shielded containers (separated by isotope, physical form [dry/liquid], and chemical form [aqueous/organic solvent phase]). Always consult your safety office for further guidance in the appropriate disposal of radioactive materials.

- Among the experiments requiring **special precautions** are those that use [^{35}S] methionine and ^{125}I, because of the dangers of airborne radioactivity. [^{35}S] methionine decomposes during storage into sulfoxide gases, which are released when the vial is opened. The isotope ^{125}I accumulates in the thyroid and is a potential health hazard. ^{125}I is used for the preparation of Bolton–Hunter reagent to radioiodinate proteins. Consult your local safety office for further guidance in the appropriate use and disposal of these radioactive materials before initiating any experiments. Wear appropriate gloves when handling potentially volatile radioactive substances, and work only in a radioiodine fume hood.

BIOLOGICAL SAFETY PROCEDURES

Biological safety fulfills three purposes: to avoid contamination of your biological sample with other species; to avoid exposure of the researcher to the sample; and to avoid release of living material into the environment. Biological safety begins with the receipt of the living sample; continues with its storage, handling, and propagation; and ends only with the proper disposal of all contaminated materials. A catalog of operations known as "sterile handling" is usually employed in manipulating living matter. However, the actual manner of treatment largely depends on the actual sample, which can be quite diverse: *Escherichia coli* and other bacterial strains, yeasts, tissues of animal or plant origin, cultures of mammalian cells, or even derivatives from human blood are routinely handled in a biological laboratory. Two of these, bacteria and human blood products, are discussed in more detail below.

The Department of Health, Education, and Welfare (HEW) has classified various bacteria into different categories with regard to shipping requirements (see Sanderson and Zeigler 1991). Nonpathogenic strains of *E. coli* (such as K12) and *Bacillus subtilis* are in Class 1 and are considered to present no or minimal hazard under normal shipping conditions. However, *Salmonella*, *Haemophilus*, and certain strains of *Streptomyces* and *Pseudomonas* are in Class 2. Class 2 bacteria are "Agents of ordinary potential hazard: agents which produce disease of varying degrees of severity... but which are contained by ordinary laboratory techniques." Contact your institution's safety office concerning shipping biological material.

Human blood, blood products, and tissues may contain occult infectious materials such as hepatitis B virus and human immunodeficiency virus (HIV) that may result in laboratory-acquired infections. Investigators working with lymphoblast cell lines transformed by Epstein–Barr virus (EBV) are also at risk of EBV infection. Any human blood, blood products, or tissues should be considered a biohazard and should be handled accordingly until proved otherwise. Wear appropriate disposable gloves, use mechanical pipetting devices, work in a biological safety cabinet, protect against the possibility of aerosol generation, and disinfect all waste materials before disposal. Autoclave contaminated plasticware before disposal; autoclave contaminated liquids or treat with bleach (10% [v/v] final concentration) for at least 30 minutes before disposal (this is also valid for used bacterial media).

Always consult your local institutional safety officer for specific handling and disposal procedures of your samples. Further information can be found in the Frequently Asked Questions of the ATCC homepage (http://www.atcc.org) and is also available from the National Institute of Environmental Health and Human Services, Biological Safety (http://www.niehs.nih.gov/about/stewardship).

GENERAL PROPERTIES OF COMMON HAZARDOUS CHEMICALS

The hazardous materials list can be summarized in the following categories.

- **Inorganic acids**, such as hydrochloric, sulfuric, nitric, or phosphoric, are colorless liquids with stinging vapors. Avoid spills on skin or clothing. Spills should be diluted with large amounts of water. The concentrated forms of these acids can destroy paper, textiles, and skin and cause serious injury to the eyes.

- **Inorganic bases**, such as sodium hydroxide, are white solids that dissolve in water and under heat development. Concentrated solutions will slowly dissolve skin and even fingernails.

- **Salts of heavy metals** are usually colored, powdered solids that dissolve in water. Many of them are potent enzyme inhibitors and therefore toxic to humans and the environment (e.g., fish and algae).

- Most **organic solvents** are flammable volatile liquids. Avoid breathing the vapors, which can cause nausea or dizziness. Also avoid skin contact.

- Other **organic compounds** including organosulfur compounds, such as mercaptoethanol or organic amines, can have very unpleasant odors. Others are highly reactive and should be handled with appropriate care.

- If improperly handled, **dyes and their solutions** can stain not only your sample but also your skin and clothing. Some are also mutagenic (e.g., ethidium bromide), carcinogenic, and toxic.

- **Nearly all names ending with "ase"** (e.g., catalase, β-glucuronidase, or zymolyase) refer to enzymes. There are also other enzymes with nonsystematic names such as pepsin. Many of them are provided by manufacturers in preparations containing buffering substances, etc. Be aware of the individual properties of materials contained in these substances.

- **Toxic compounds** are often used to manipulate cells. They can be dangerous and should be handled appropriately.

- **Be aware** that several of the compounds listed have not been thoroughly studied with respect to their toxicological properties. Handle each chemical with appropriate respect. Although the toxic effects of a compound can be quantified (e.g., LD_{50} values), this is not possible for carcinogens or mutagens where one single exposure can have an effect. Also realize that dangers related to a given compound may also depend on its physical state (fine powder vs. large crystals/diethyl ether vs. glycerol/dry ice vs. carbon dioxide under pressure in a gas bomb). Anticipate under which circumstances during an experiment exposure is most likely to occur and how best to protect yourself and your environment.

Cold Spring Harbor Laboratory Press (CSHLP) has used its best efforts in collecting and preparing the material contained herein but does not assume, and hereby disclaims, any liability for any loss or damage caused by errors and omissions in the publication, whether such errors and omissions result from negligence, accident, or any other cause. CSHLP does not assume responsibility for the user's failure to consult more complete information regarding the hazardous substances listed in this publication.

REFERENCE

Sanderson KE, Zeigler DR. 1991. Storing, shipping, and maintaining records on bacterial strains. *Methods Enzymol* **204**: 248–264.

WWW RESOURCES

ATCC Home page http://www.atcc.org

ATCC, for Sample Handling (in Frequently Asked Questions) http://www.atcc.org/Culturesand Products/TechnicalSupport/FrequentlyAskedQuestions/tabid/469/Default.aspx

GraphPad Software, Radioactivity Calculations http://www.graphpad.com/quickcalcs/Chem Menu.cfm

National Institute of Environmental Health and Human Services, Biological Safety (NIEHS) http://www.niehs.nih.gov/about/stewardship

U.S. Environmental Protection Agency (EPA), Federal waste disposal regulations, Laboratory http://www.epa.gov/epawaste/hazard/tsd/index.htm

U.S. Environmental Protection Agency (EPA), Individual States and Territories http://www.epa. gov/epawaste/wyl/stateprograms.htm

U.S. Nuclear Regulatory Commission (NRC), Medical Pathological Radioactively Contaminated Waste (Decay-in-Storage) http://www.nrc.gov/reading-rm/doc-collections/cfr/part035/ part035-0092.html

U.S. Nuclear Regulatory Commission (NRC), Radioactive Waste Disposal Regulations: General Requirements http://www.nrc.gov/reading-rm/doc-collections/cfr/part020/part020-2001. html

Index

A

α-Aminoadipate plates
(α-AA), 202–203
Acetamide media, 207
Actin staining in fixed yeast
cells
materials and solutions,
156
procedure, 155–156
safety notes, 155
Add-back medium, 196–197
ADE2, 2, 91
ade2 mutants, 19
Aequoria victoria, 20
Affymetrix microarrays, 113
Agarose pads, 157–158
AID (auxin-inducible degron)
system, 70
ARS (autonomous replicating
sequence), 2–3
Ascospores, 43
Ascus, 43
Ashbya gossypii, 124
ATCC, 175, 223
Autonomous replicating
sequence (*ARS*), 2–3
Auxin-inducible degron
(AID) system, 70
Auxotrophic mutants,
45, 49, 57

B

Barcode sequencing
about, 113–114
experimental procedures,
114–121
bar-seq PCR, 116–119
colony PCR, 119–121

guarding against
contamination, 114
materials, 121–122
strains, 114
Bar-seq PCR, 116–119
Basal copy number, 98
Bialaphos, 206
Biological safety procedures,
222–223
Bulk segregant analysis, 99
BY4743 background, 79

C

C6 cytometer
adding solution, 140–141
after sampling, 140
general procedures, 138
running the samples,
139–140
start-up, 138–139
vocabulary, 137–138
$CaCl_2$, 1
Calcofluor, 17, 18*f*, 154
CAN1 gene, 90
Canavanine, 81*f*, 82, 83, 90,
92, 94–95, 204,
207–208
Cassette models, 30, 30*f*
Centromere (CEN)
plasmids, 3
CFP (cyan fluorescent
protein), 20
CGH. *See* Comparative
genomic hybridization
Chemostat, 105
cis-trans test. *See*
Complementation
testing
clonNAT, 205

CNVs (copy-number
variations), 97–98
Colony PCR, 119–121, 191
Colony pinning, 83
Comparative functional
genomics. *See* Barcode
sequencing
Comparative genomic
hybridization (CGH)
copy-number changes and,
97–98
experimental procedures,
100–103
following up candidate
mutations, 99–100
linking mutations to
phenotype, 98–99
materials, 103–104
strains, 100
Complementation testing,
58–59
Conditional expression
systems, 70–72
Copy-number variations
(CNVs), 97–98
Counting yeast cells, 93–94,
167–169
Cyan fluorescent protein
(CFP), 20
Cycloheximide, 203
Cytometer. *See* C6 cytometer

D

DAPI (4′,6-Diamidino-2-
phenylindole), 17–18,
18*f*, 145, 151–153
Deletion collection strains
storage and handling,
175–177

Deletion Consortium, 175
Deletion mutations
    classes of genes defined
        by, 79
    experimental procedures,
        84–86
    materials, 87
    phenotype screening, 80
    random spore analysis
        and, 82–84
    strains, 84
    strains vendors, 79
    synthetic-lethal interactions
        identification, 80–82,
        81f, 83–84
    uses for in yeast genetics, 80
Department of Health,
        Education, and
        Welfare (HEW), 223
4′,6-Diamidino-2-
        phenylindole (DAPI),
        17–18, 18f, 145,
        151–153
DiOC₆(3), 18, 153
Dihydrofolate reductase
        (DHFR), 70
DiIC₅(3), 18, 153
Diploids
    cell morphology and, 16,
        16f, 98
    complementation testing
        and, 59
    haploid strains and, 98
    mating-type and, 29, 30,
        31–32, 31f
    meiosis and, 43
    selection using SGA, 81f,
        82–83
    tetrad dissection and, 45, 49
    transformation practices,
        124
Discosoma, 20
Disposal of materials. See
        Safety and hazardous
        material information
Dissection needles, making,
        163–164
DNA
    barcode sequencing and,
        113–114
    concentration
        measurement, 189

copy-number
    measurement, 97–98
DAPI and, 17–18
essential genes and, 69
genomic modifications with
    PCR products and, 123,
    126f, 128, 129
mutation detection and,
    108–109
preparation of genomic,
    143–144
staining of, 151
uptag and downtag, 176
yeast transformation
    methods and, 1–8
Dominant drug resistance, 2
Downtagging, 113, 175,
    176
Dropout media, 197–199
Drug selection media
    5-fluoroanthranilic acid,
        203
    5-fluoro-orotic acid plates,
        201–202
    α-aminoadipate plates,
        202–203
    acetamide media, 207
    bialaphos, 206
    canavanine, 204
    clonNAT, 205
    cycloheximide, 203
    fluoroacetamide/
        acetamide, 206–207
    5-fluoroanthranilic acid,
        203
    5-fluoro-orotic acid plates,
        201–202
    G418, 204–205
    hygromycin, 205
    phleomycin/zeocin, 206
    thialysine, 204

E

Electroporation, 1–2
EMS (ethylmethane
    sulfonate), 57, 80,
    165–166
EMS mutagenesis
    materials and solutions,
        166
    procedure, 165–166

safety notes, 165
Epigenetic transcriptional
    silencing
    assays to monitor mating-
        type silencing, 31–33,
        31f, 32f
    experimental procedures,
        35–40
    experiment overview, 34
    HML and HMR mating loci,
        30, 30f
    HML and HMR silencing,
        30–31
    materials, 40–41
    mating-type and, 29–30,
        30f, 31f
    plasmids, 34
    practical applications of
        mating yeast, 33
    strains, 33–34
    variegated expression and
        epigenetic
        inheritance, 31
Essential genes study
    conditional expression
        systems, 70–72
    examples of essential
        genes, 69
    experimental procedures,
        73–75
    materials, 76–77
    point mutants, 70
    strains, 72–73
    temperature-sensitive
        mutants, 69–70
Ethylmethane sulfonate
    (EMS), 57, 80,
    165–166
European Saccharomyces
    cerevisiae Archive for
    Functional Analysis
    (EUROSCARF), 79,
    123, 175

F

Fiber-optic needles, 163–164
Flow cytometry of yeast DNA
    content
    materials, 172–173
    procedure, 171–172
Fluorescence resonance energy
    transfer (FRET), 20

Fluoroacetamide/acetamide, 206–207
5-Fluoroanthranilic acid (5-FAA), 203
FM 4–64, 18, 18f, 153
5-FOA (5-fluoro-orotic acid), 6, 32, 57, 201–202

G

G418, 204–205
GAL1 system, 70
Gal4 transcriptional activator, 33
GE Healthcare Dharmacon, 79, 175
Genomic modifications with PCR products
  availability of PCR templates, 123
  de novo gene disruption by one-step gene replacement, 124–125
  gene disruption by one-step gene replacement, 125–126
  generation of protein fusions by one-step modification, 127–128
  mistargeting and, 123–124
  PCR protocol for gene modifications, 128–130
  reagents, 130
  single-step attribute, 123
  transforming diploid yeast strains first and, 124
GFP (green fluorescent protein), 20
Glass needles, 163

H

Haploids
  cell morphology and, 15, 16, 16f, 17, 98
  diploid strains and, 98
  mating-type and, 29, 30, 31–32, 31f
  meiosis and, 43–44
  selection using SGA, 81f, 82–83
  tetrad dissection and, 45–46, 48–49

transformation practices, 124
Hazardous chemicals, 223–224
Hemocytometer, 167–169
Heterothallic haploid strains, 29
HEW (Department of Health, Education, and Welfare), 223
High-efficiency yeast transformation
  materials, 135
  procedure, 133–134
HIS3, 2
HML and HMR
  mating loci, 30, 30f
  silencing, 30–31
Hoffman–Winston genomic DNA preparation, modified
  materials, 144
  procedure, 143–144
  safety notes, 143
HO gene, 29
Homothallic haploid strains, 29
Hygromycin, 205

I

Imaging of live yeast, 157–158
Indicator media
  limiting Ade plates, 200
  MAL, 201
  X-Gal plates, 201
Indirect immunofluorescence
  about, 20–21
  materials and solutions, 148–150
  microscopy procedure
    antibody staining, 147–148
    cell fixation, 145
    mounting, 148
    optional DAPI staining, 148
    slide preparation, 146–147
    spheroplasting, 145–146
  safety notes, 145
Institutional safety officer, 217

K

Karyogamy, 17
Kluyveromyces lactis, 124

L

Laboratory waste disposal, 220–221
LEU2, 2
Light microscopy
  examination of growth properties, 15–16, 16f
  experimental procedures, 23–26
  experiment overview, 22–23
  with fluorescent protein chimeras, 20
  haploids versus diploids, 16, 16f, 17
  indirect immunofluorescence, 20–21
  materials, 26–28
  mating cells, 17
  mutants behavior, 19
  plasmids, 22
  primers, 22
  safety notes, 22
  staining cells with dyes and drugs, 17–19, 18f
  strains, 21–22
Limiting Ade plates, 200
Linkage analysis, 99
Lithium acetate (LiOAc), 1

M

MAL indicator plates, 201
Material safety data sheets (MSDSs), 217
Mating, meiosis, and tetrad dissection
  experimental procedures, 49–54
  materials, 54–55
  meiosis of MATa/MATα diploids, 43, 44f
  strains, 49
  tetrad analysis, 45, 47–48, 48f

Mating, meiosis, and tetrad
dissection (*Continued*)
tetrad dissection, 43–44,
44*f*, 48–49
types of tetrads, 45–47,
46*f*, 48*f*
Mating-type
about, 29–30, 30*f*, 31*f*
alleles, 29
assays to monitor
mating-type silencing,
31–33, 31*f*, 32*f*
MAT locus, 29, 30
mCherry, 20, 127
Measuring DNA
concentration, 189
Media
avoiding caramelization,
195
avoiding mushy plates, 195
drug selection, 201–207
indicator, 200–201
rich, 195–196
SGA, 207–209
sporulation, 199–200
synthetic, 196–199
Medial pathological waste
(MPW), 220
Meiosis. *See* Mating, meiosis,
and tetrad dissection
MET3 system, 70
Microscopy. *See* Light
microscopy
Minimal SPO, 199
Mitochondrial dyes, 18–19,
18*f*
Mitotracker Red, 18, 18*f*,
19, 153
Modified Hoffman–Winston
genomic DNA
preparation
materials, 144
procedure, 143–144
safety notes, 143
MPW (medial pathological
waste), 220
MSDSs (material safety data
sheets), 217
MSH2, 90
2μ-Based plasmids, 3
Mutant isolation and
characterization

auxotrophic mutants, 57
complementation testing,
58–59
experimental procedures,
60–65
experiment overview, 59
materials, 65–66
mutant enrichment, 57–58
safety notes, 60
strains, 59–60
temperature-sensitive
mutants, 58
Mutation rates
cell counting, 93–94
experimental approach,
90–91
experimental procedures,
91–95
growth equation, 89
materials, 95–96
measurement technique,
89–90
plating mutants, 94–95
qualitative view of, 91
strains, 91
Mutations
deletion (*see* Deletion
mutations)
detecting using CGH
(*see* Comparative
genomic
hybridization)
detecting using the
whole genome (*see*
Whole-genome
sequencing and
linkage)
isolation and
characterization (*see*
Mutant isolation and
characterization)
rates (*see* Mutation rates)

N

Nalgene, 83
National Institute of
Environmental Health
and Human Services
(NIEHS), 223
N-degron system, 70
Needles, making, 163–164

Nonparental ditype (NPD)
tetrad, 45
Nystatin, 58

O

OmniTrays, 83

P

Paraformaldehyde, 145, 155
Parental ditype (PD) tetrad, 45
PCR (polymerase chain
reaction)
amplification, 109–110
barcode sequencing
and, 114
bar-seq procedure,
116–119
clean-up, 110–111
colony, 119–121, 191
genomic modifications
using
availability of PCR
templates, 123
de novo gene disruption
by one-step gene
replacement, 124–125
gene disruption by
one-step gene
replacement, 125–126
generation of protein
fusions by one-step
modification, 127–128
mistargeting and,
123–124
PCR protocol for gene
modifications,
128–130
reagents, 130
single-step attribute, 123
transforming diploid
yeast strains first
and, 124
integration analysis using,
1, 5
isolating mutants and, 70
mediated genome
modifications using,
6–7
mutation detection and, 98
plasmids transformation
and, 6–7, 6*f*

PEG (polyethylene glycol), 1
Petite phenotype, 19
PFGE (pulsed-field gel electrophoresis), 98
Phenol, 143
Phenylenediamine, 145
Pheromones and mating cells, 17
Phleomycin/zeocin, 206
Plasmids transformation
  ARS elements and replication, 2–3
  experimental procedures, 8–11
  experiment overview, 7–8
  integration into the yeast genome, 4–6, 5f
  integration reversal, 5–6
  materials, 11–13
  PCR-mediated genome modifications, 6–7, 6f
  plasmids, 7
  primers, 7
  replication origins, 2–3
  selectable markers, 2
  strains, 7
  transformation methods for yeast, 1–2
  vector systems, 3–4, 4f
  verification of integration at a specific locus, 4–5
PlusPlates, 83
Point mutants, 70
Polyethylene glycol (PEG), 1
Polymerase chain reaction. *See* PCR
Pop-in/Pop-out technique, 5–6
Presporulation media, 200
pRS406-Nup49-GFP, 4
pRS series, 3
Pulsed-field gel electrophoresis (PFGE), 98

Q

Qubit fluorometer, 189

R

Radioactive safety procedures, 221–222

Random spore analysis, 82–84
Red fluorescent protein (RFP), 20
Rhodamine-phalloidin, 18f, 19
Rich media
  YEP, 196
  YPD (YEPD), 195
  YPG (YEPG, YEP-glycerol), 196

S

S&P Robotics, 177
Sac, 43
Saccharomyces cerevisiae
  deletion mutations, 79–84
  epigenetic transcriptional silencing, 29–34
  essential genes study, 69–72
  growth properties, 15–16, 16f
  haploids versus diploids, 16, 16f, 17
  light microscopy use in research, 15–22
  mating, meiosis, and tetrad dissection, 43–49
  mating cells, 17
  mating-type, 29, 29–30, 30f, 31f
  mutant isolation and characterization, 57–59
  mutants behavior, 19
  mutation detection using CGH, 97–100
  mutation detection using whole-genome sequencing, 105–112
  mutation rates, 89–91
  plasmids transformation, 1–8
  tetrads, 43–49
Saccharomyces Genome Database (SGD), 47, 175
Safety and hazardous material information
  biological safety procedures, 222–223
  classes of agents, 223

general safety and disposal cautions, 217, 218–220
hazardous chemicals
  general properties, 223–224
laboratory waste disposal, 220–221
primary safety information resources, 217–218
radioactive safety procedures, 221–222
Safety notes. *See also* Safety and hazardous material information
  actin staining in fixed yeast cells, 155
  EMS mutagenesis, 165
  Hoffman–Winston genomic DNA preparation, 143
  indirect immunofluorescence, 145
  for light microscopy, 22
  mutant isolation and characterization, 60
Sanger sequencing, 99
Schizosaccharomyces pombe, 124
Scoring SGA screens. *See* SGATools
Selectable markers, 2
SGA (synthetic genetic array), 80–82, 81f, 83–84
SGA media
  amino acids mix, 209
  SD/MSG -HIS -ARG -LYS mixes, 207–208
  SPO+YE+glucose+amino acids, 207
SGATools, 83
  analyzing screens
    calculating normalized colony sizes, 183–185
    displaying genetic interaction scores, 185–188
    obtaining colony sizes, 180–182
    ORF names, 180
    process images and obtaining colony sizes, 180–182

SGATools (*Continued*)
  scoring genetic
    interactions, 183–185
  quantitative scoring of
    genetic interactions,
    179
SGD (*Saccharomyces* Genome
    Database), 47, 175
Shmooing cells, 17
Sigma-Aldrich, 198
Singer Instruments, 177
Singer RoToR colony pinning
    robot, 83
Single-nucleotide variants
    (SNVs), 98
Sir1/2/3/4, 31
Slide preparation, 157–158
Spheroplasting, 1, 2
SPO, 199
Sporulation
  defined, 43
  medium
    minimal SPO, 199
    presporulation, 200
    SPO, 199
  tetrad dissection and
    liquid sporulation
      protocol, 159–160
    materials, 162
    plate sporulation
      protocol, 160
    preparation of tetrads,
      160–162
Staining cells
  actin staining in fixed yeast
    cells, 155–156
  with dyes and drugs, 17–19,
    18*f*
  yeast vital stains,
    151–154
Streaks and patches
    templates, 215–216
Sunrise Science, 198
Supplemented minimal
    medium (SMM),
    196–197
SYBR green, 171–173
Synthetic complete (SD)
    media, 197–199
Synthetic dextrose minimal
    medium (SD), 196

Synthetic genetic array
    (SGA), 80–82, 81*f*,
    83–84
Synthetic–lethal
    interactions, 80–82,
    81*f*, 83–84
Synthetic media
  add-back medium,
    196–197
  dropout, 197–199
  supplemented minimal
    medium, 196–197
  synthetic complete,
    197–199
  synthetic dextrose, 196
Systemic deletion collection
    storage and handling,
    175–177

T

Teknova, 198
Temperature-sensitive
    mutants, 58, 69–70
Templates
  96-Well plate, 213
  for making streaks and
    patches, 215–216
Tet-off system, 70
Tetrads
  analysis, 45, 47–48, 48*f*
  dissection, 43–44, 44*f*,
    48–49
  dissection and sporulation
    liquid sporulation
      protocol, 159–160
    materials, 162
    plate sporulation
      protocol, 160
    preparation of tetrads,
      160–162
  dissection sheets,
    211–212
  experimental procedures,
    49–54
  making a dissection needle,
    163–164
  materials, 54–55
  strains, 49
  types of
    centromere-linked
      genes, 47
    classes of, 45, 46*f*, 48*f*

linkage relationships
    possible, 45
  same chromosome, less
    tightly linked, 46
  same chromosome,
    tightly linked, 45–46
  unlinked genes, 47
Tetratype (T) tetrad, 45
Thialysine, 82, 83, 204
Tiling microarrays, 99
Training for the plate
    race, 193
transOMIC technologies,
    79, 175
TRP1, 2

U

Ubiquitin, 70
Uptagging, 113, 175, 176
URA3, 2, 4, 4*f*, 5, 32, 32*f*, 57
U.S. Environmental
    Protection Agency,
    220
U.S. Nuclear Regulatory
    Commission, 220

V

V&P Scientific, 83, 176
Vector systems, 3–4, 4*f*
Vital stains
  calcofluor staining of
    chitin and bud
    scars, 154
  DAPI staining of nuclear
    and mitochondrial
    DNA
    materials, 152–153
    procedures, 151–152
    safety notes, 151
  visualization of
    mitochondria, 153
    vacuoles and endocytic
      compartments, 154

W

96-Well plate template,
    213
Whole-genome sequencing
    and linkage

experimental procedures
  clean-up of tagmented
    DNA with Zymo
    columns, 109
  DNA quantification and
    quality, 108
  DNA tagmentation, 108
  mating-type testing,
    111–112
  PCR amplification,
    109–110
  PCR clean-up, 110–111
  plate preparation,
    106–107
  materials, 112
  strains, 105

X

X-Gal plates, 201

Y

Yeast transformation,
    high-efficiency
  materials, 135
  procedure, 133–134
Yellow fluorescent protein
    (YFP), 20
YEP, 196
YPD (YEPD), 195
YPG (YEPG, YEP-glycerol),
    196

Z

Zygote, 17

www.ingramcontent.com/pod-product-compliance
Lightning Source LLC
Chambersburg PA
CBHW050344230326

41458CB00101B/6312